# Biosystems & Biorobotics

Volume 18

*Aims & Scope*

Biosystems & Biorobotics publishes the latest research developments in three main areas: 1) understanding biological systems from a bioengineering point of view, i.e. the study of biosystems by exploiting engineering methods and tools to unveil their functioning principles and unrivalled performance; 2) design and development of biologically inspired machines and systems to be used for different purposes and in a variety of application contexts. The series welcomes contributions on novel design approaches, methods and tools as well as case studies on specific bioinspired systems; 3) design and developments of nano-, micro-, macrodevices and systems for biomedical applications, i.e. technologies that can improve modern healthcare and welfare by enabling novel solutions for prevention, diagnosis, surgery, prosthetics, rehabilitation and independent living.

On one side, the series focuses on recent methods and technologies which allow multiscale, multi-physics, high-resolution analysis and modeling of biological systems. A special emphasis on this side is given to the use of mechatronic and robotic systems as a tool for basic research in biology. On the other side, the series authoritatively reports on current theoretical and experimental challenges and developments related to the "biomechatronic" design of novel biorobotic machines. A special emphasis on this side is given to human-machine interaction and interfacing, and also to the ethical and social implications of this emerging research area, as key challenges for the acceptability and sustainability of biorobotics technology.

The main target of the series are engineers interested in biology and medicine, and specifically bioengineers and bioroboticists. Volume published in the series comprise monographs, edited volumes, lecture notes, as well as selected conference proceedings and PhD theses. The series also publishes books purposely devoted to support education in bioengineering, biomedical engineering, biomechatronics and biorobotics at graduate and post-graduate levels.

*About the Cover*

The cover of the book series Biosystems & Biorobotics features a robotic hand prosthesis. This looks like a natural hand and is ready to be implanted on a human amputee to help them recover their physical capabilities. This picture was chosen to represent a variety of concepts and disciplines: from the understanding of biological systems to biomechatronics, bioinspiration and biomimetics; and from the concept of human-robot and human-machine interaction to the use of robots and, more generally, of engineering techniques for biological research and in healthcare. The picture also points to the social impact of bioengineering research and to its potential for improving human health and the quality of life of all individuals, including those with special needs. The picture was taken during the LIFEHAND experimental trials run at Università Campus Bio-Medico of Rome (Italy) in 2008. The LIFEHAND project tested the ability of an amputee patient to control the Cyberhand, a robotic prosthesis developed at Scuola Superiore Sant'Anna in Pisa (Italy), using the tf-LIFE electrodes developed at the Fraunhofer Institute for Biomedical Engineering (IBMT, Germany), which were implanted in the patient's arm. The implanted tf-LIFE electrodes were shown to enable bidirectional communication (from brain to hand and vice versa) between the brain and the Cyberhand. As a result, the patient was able to control complex movements of the prosthesis, while receiving sensory feedback in the form of direct neurostimulation. For more information please visit http://www.biorobotics.it or contact the Series Editor.

More information about this series at http://www.springer.com/series/10421

Gianluca Borghini · Pietro Aricò
Gianluca Di Flumeri · Fabio Babiloni

# Industrial Neuroscience in Aviation

## Evaluation of Mental States in Aviation Personnel

Springer

Gianluca Borghini
BrainSigns
Sapienza University of Rome
Rome
Italy

Gianluca Di Flumeri
BrainSigns
Sapienza University of Rome
Rome
Italy

Pietro Aricò
BrainSigns
Sapienza University of Rome
Rome
Italy

Fabio Babiloni
BrainSigns
Sapienza University of Rome
Rome
Italy

ISSN 2195-3562 ISSN 2195-3570 (electronic)
Biosystems & Biorobotics
ISBN 978-3-319-58597-0 ISBN 978-3-319-58598-7 (eBook)
DOI 10.1007/978-3-319-58598-7

Library of Congress Control Number: 2017939608



Printed on acid-free paper

This Springer imprint is published by Springer Nature
The registered company is Springer International Publishing AG
The registered company address is: Gewerbestrasse 11, 6330 Cham, Switzerland

# Preface

*Industrial Neuroscience* refers to an emerging area of science where state-of-the-art methodologies are applied in real working contexts, in order to evaluate the internal cognitive and emotional state of human operators. In particular, aviation is one of the industrial contexts that mainly benefit from the use of neuroscience's knowledge and technologies. With the generic term 'aviation' is indicated here as an area of research that includes both the study of personnel involved in piloting aircrafts or helicopters, and the personnel involved in managing the air traffic, i.e., Air Traffic Controllers. The importance of using neuroscience in aviation relies on the fact that nowadays several studies have demonstrated that the main cause of accidents in aviation are due to *Human Factors* (HF). This happens because humans are the least and last controllable factor in the activity's workflows, and the availability of tools able to provide objective information about the user's cognitive state should be very helpful in maintaining proper levels of safety and improve human–machine interactions. To proper illustrate how applied neuroscience could help in this process, the book will cover three topics—all of them intrinsically connected. The first one is focused on the most important and investigated mental states in aviation. In particular, state-of the-art descriptions of the mental workload, cognitive control behavior, and training will be provided. The second topic of the book is dedicated to the experimental environments in which the presented research activities have been performed, the methodologies and machine learning-based algorithm developed to objectively assess the considered operator's mental states along the execution of tasks. In particular, the developed algorithm has been studied to address two important limitations: (i) overtime reliability (no recalibration of the algorithm); (ii) automatic brain features selection to avoid both the *underfitting* and *overfitting* problems. In the final part of the book, an overview of the results will be provided. The experiments have been initially run in controlled environments involving voluntary students, whilst the final tests have been carried out in realistic environments recruiting professional personnel.

Rome, Italy

Gianluca Borghini
Pietro Aricò
Gianluca Di Flumeri
Fabio Babiloni

# Contents

# Abbreviations

| | |
|---|---|
| AA | Adaptive Automation |
| ACC | Anterior Cingulate Cortex |
| ACC | Area Control Center |
| AI | Artificial Intelligence |
| AIS | Aeronautical Information Services |
| ANOVA | Analysis of Variance |
| ANS | Autonomic Nervous System |
| asSWLDA | Automatic Stop StepWise Linear Discriminant Analysis |
| ATC | Air Traffic Controller |
| ATCO | Air Traffic Controller Officer |
| ATFM | Air Traffic Flow Management |
| ATM | Air Traffic Management |
| ATP | Adenosine Triphosphate |
| AUC | Area Under Curve |
| AUGCOG | Augmented Cognition |
| BA | Brodmann Area |
| BCI | Brain Computer Interface |
| BOLD | Blood Oxygen Level Dependent |
| CC | Cortico Cerebellar |
| CI | Confidence Interval |
| Class-ndx | Classification index |
| Class-THR | Classification Threshold |
| CMA | Cingulate Motor Area |
| CNS | Central Nervous System |
| COMM | Auditory Monitoring |
| CS | Cortico Striatal |
| CWP | Controller Working Position |
| DARPA | Defense Advanced Research Projects Agency |
| DLPFC | Dorsolateral Prefrontal Cortex |
| DM | Decision-Making |

| | |
|---|---|
| EBR | Eye Blinks Rate |
| ECG | Electrocardiography |
| ECoG | Electrocorticography |
| EEG | Electroencephalogram |
| ENAC | École Nationale de l'Aviation Civile |
| ENAV | Ente Nazionale di Assistenza al Volo |
| EOG | Electrooculogram |
| EPSP | Excitatory or Depolarizing Postsynaptic Potential |
| FDR | False Discovery Rate |
| FEF | Frontal Eye Field |
| FFT | Fast Fourier Transform |
| FL | Flight Level |
| fMRI | Functional Magnetic Resonance Imaging |
| fNIR | Functional Near InfraRed |
| FOS | Fast Optical Signal |
| GABA | Gamma Amino Butyric Acid |
| GSR | Galvanic Skin Response |
| HDD | Head Down Display |
| HF | Human Factor |
| HMD | Head Mounted Display |
| HMI | Human Machine Interaction |
| HR | Heart Rate |
| HUD | Head Up Display |
| IAF | Individual Alpha Frequency |
| ICA | Independent Component Analysis |
| IPS | Intraparietal Sulcus |
| IPSP | Inhibitory or Hyperpolarizing Postsynaptic Potential |
| ISA | Instantaneous Self Assessment |
| LDA | Linear Discriminant Analysis |
| LT | Laplace Transform |
| MA-Class | Moving Average Classification index |
| MATB | Multi Attribute Task Battery |
| MEG | Magneticencephalogram |
| mPFC | Medial Prefrontal Cortex |
| MWL | Mental Work Load |
| NASA-TLX | NASA Task Load Index |
| NINA | Neurometric Indicators for ATM |
| NN | Neural Networks |
| NUS | National University of Singapore |
| OFC | Orbitofrontal Cortex |
| p-BCI | Passive Brain Computer Interface |
| PET | Positron Emission Tomography |
| PFC | Prefrontal Cortex |
| PNS | Peripheral Nervous System |
| PPC | Posterior Parietal Cortex |

| | |
|---|---|
| PSD | Power Spectral Density |
| RMAN | Resource Management task |
| ROC | Receiver Operating Characteristic |
| ROI | Region of Interest |
| RT | Reaction Times |
| SRK | Skill, Rule and Knowledge |
| SRT | Serial Reaction Time |
| STCA | Short Term Conflict Alert |
| SVM | Support Vector Machines |
| SWLDA | Stepwise Linear Discriminant Analysis |
| SYSM | Response to Event onsets task |
| tfICA | Time–Frequency Independent Component Analysis |
| TMS | Transcranial Magnetic Stimulation |
| TP | Task Performance |
| TPJ | Temporoparietal Junction |
| TRCK | Tracking Task |
| TTB | Take-The-Best |
| UAV | Unmanned Aerial Vehicles |
| VLPFC | Ventrolateral Prefrontal Cortex |
| VMPFC | Ventromedial Prefrontal Cortex |
| WEEG | EEG-based Mental Workload Index |
| WL | Work Load |
| WM | Working Memory |
| WMRT | Weighted Mean Reaction Time |

# List of Tables

# Chapter 1
# Introduction

In operational environments, safety of people relies on the work and efficiency of one or more operators: in such contexts, a human error could have serious and dramatic consequences. For example, in the transports' domain, the passengers' safety depends on the performance of the Pilot(s), on the Air Traffic Controller(s), or on the Driver(s). In general, *Human Factors* (HFs) have consistently been identified as one of the main factors—in a high proportion—of all workplaces accidents. In particular, it has been estimated that up to 90% of accidents exhibits HF as principal cause (Feyer and Williamson 2011). For example, in the domain of the health care, the US Institute of Medicine estimates that there is a high people mortality per year (between 44,000 and 88,000) as a result of medical errors (Helmreich 2000), and it has also been estimated that inappropriate human actions are implicated in as much as 95% of road traffic crashes (Aberg and Rimmö 1998). Additionally, over the past four decades, HF has been involved in a high number of casualty catastrophes (Salmon et al. 2005). Consequently, the HF construct has received more attention and has been investigated across a wide range of domains, including military and civil aviation (Shappel and Wiegmann 2000; Stanton et al. 2006), aviation maintenance (Rankin et al. 2000), air traffic management (Shorrock and Kirwan 2002), rail (Lawton and Ward 2005), road transport (Reason 2016; Rumar 1990), nuclear power and petrochemical reprocessing (Kirwan 1998; Sexton et al. 2000; Shorrock and Kirwan 2002), military medicine (Helmreich 2000; Sexton et al. 2000), and even the space travel domain (Nelson et al. 1998).

Human error is an extremely common phenomenon, since people, regardless of their abilities, skill level, and expertise, make mistakes. The typical consequence of error occurrence is the failure to achieve a desired outcome or the production of an undesirable outcome. When it happens in particular working environments, such error can potentially lead to accidents involving injury and fatalities. According to the scientific literature, there have been numerous attempts at defining and classifying the human error. However, a universally accepted definition does not exist yet. Rasmussen (1982) pointed out the difficulty in providing a satisfactory definition of human error. In (Rasmussen 1987), he suggested that human error



G. Borghini et al., *Industrial Neuroscience in Aviation*,
Biosystems & Biorobotics 18, DOI 10.1007/978-3-319-58598-7_1

represents a mismatch between the demands of an operational system and what the operator does. The main causes of human errors have to be sought within the internal or psychological factors of the operator (Reason 2008, 2016). In fact, errors could arise from aberrant mental processes, such as inattention, poor motivation, loss of vigilance, mental overload, and fatigue that negatively affect the user's performance. For example, cognitive psychology literature demonstrated that mental workload has an "inverted U-shape" relationship with performance. In other words, some levels of mental workload may help the user to reach high levels of performance (Calabrese 2008; Warm et al. 2008), since it positively stimulates the user and keeps him/her awake with high attention levels. On the contrary, a period of mental inactivity and "under-stimulation" can cause a monotonous and boring state (*underload*), a low level of vigilance and attention, with a low demand of cognitive resources. Additionally, an operative condition characterized by high demanding multitasks can lead the user to an *overload* condition and to a likely occurrence of errors (Kirsh 2000).

Both the cases bring to a variation of neurophysiological factors and often to a decrement of performance. Such performance reduction is highly undesired, especially in critical domains, as discussed above. In this regard, the *augmented cognition* research field aims at developing systems to avoid performance degradation by adapting the user's interface and reducing the task demand/complexity, or by intervening directly on the system (Fuchs et al. 2007). Over the past two decades, researchers in the field of augmented cognition developed novel technologies to both monitor and enhance human cognition and performance. Most of those works were based on evidences coming from cognitive science and cognitive neuroscience (Decades 2008). On the basis of such findings and technological improvements in measuring human biosignals, it has been possible to evaluate operators' mental states unobtrusively and in realistic contexts. The neurophysiological indexes have then been used as *inputs* for the interface the operator was interacting with to improve the *Human–Machine Interaction* (HMI). In fact, during the last decades, the rapid growth of highly complex and dynamic human–machine systems boosted a wide development of research on HMI, with the aim to deal with the ergonomic challenges introduced by such systems. In this context, HF researches have developed a large number of studies on failures related to the human–machine relationship (Giraudet et al. 2015; Parasuraman et al. 1994; Reason 2008; Wickens 1992) in order to improve and ensure good interactions between the work environment and human capabilities (Wickens 1992). From human operators' point of view, machines are not only tools, but also possibly autonomous agents. Thus, any kinds of coherence must be maintained between humans and machines' actions on the environment (Hoc 2000). Humans can adapt themselves to a variety of work environments, performing simultaneously various tasks, using different equipments (screens, touch interfaces, keyboards, cloche, etc.). Obviously, the greater the number and variety of tasks to perform and devices to use is, the higher the workload experienced is. Therefore, in recent studies it has been investigated the possibility of developing intelligent systems able to *automatically* support the operator in executing part of or entire tasks in order to reduce the experienced workload, and consequently improve the performance and limit the probability of errors commission. In general, the term "*Automation*" refers to the process of entirely or partially allocating the activities constituting a task usually performed by a human, to a machine or a system (Parsons et al. 2005). The degree to which an activity is automated is referred as *Level Of Automation* (LOA). The most comprehensive list was developed by Sheridan (Sheridan 2002; Sheridan and Verplank 1978). The LOAs can range from *complete human control* to *complete computer control*. In particular, (Sheridan 1992) identified 10 different LOAs, differing in the number and type of activities of the whole task allocated to the operator and to the system, where the Level 1 corresponds to the *operator's fully manual control*, while Level 10 to the *system fully automated control*.

Systems which automated aids are implemented dynamically, in response to changing task demands on the operator, are called *Adaptive Automation* (AA). Rouse (1976) stated that "The concept of adaptive aiding [...], emerged in 1974 in the course of an Air Force Systems Command-sponsored project at the University of Illinois that, was concerned with applications of *Artificial Intelligence* (AI) to cockpit automation."

In this context, the *Dynamic Function Allocation* (DFA) is a peculiar aspect of AA (Kaber et al. 2001). It consists in assigning the authority on specific functions to either the human operator or the automated system, depending on the overall context (i.e. operator's and environment conditions), and on a defined set of criteria. In fact, although the operator is normally able to perform a particular task, he/she may get tired after several hours of operations, or he/she may find it difficult to operate under time pressure or in critical situations when concurrent tasks have to be performed or particular events occur. At this point, it is not only a "*who does what*" problem, but "*who does what and when*" considerations are needed for the success and effectiveness of Functions Allocation (Inagaki 2003).

Several strategies concerning the DFA, i.e. the triggering mechanism for shifting among modes or levels of automation, have been proposed (Parasuraman et al. 1992; Scerbo et al. 2003). Three are the main approaches described in the literature:

1. *Critical-event strategy*: The system might invoke a specific LOA when specific events (called *critical events*) occur in the human–machine system. It is assumed that human workload may become unacceptably high when the critical events occur. Allocation of operative functions would not be altered if the critical events did not occur during the system operation. For example, AA would be invoked in an air traffic control system in occurrence of a rapid increasing in traffic density or complexity, by presenting to the controller decision aids for conflict detection and resolution (Hilburn et al. 1997).
2. *Performance-measurement strategy*: Operator's performance models, based on the operative task itself or on additional tasks (also called *behavioral measures*), can be used to estimate current and predicted operator's state and to infer whether workload is excessive or not (Scerbo et al. 2001).

3. *Neurophysiological measurement strategy*: information about the neurophysiological state of the operator can be obtained by recording his neurophysiological signals, e.g. *electroencephalogram* (EEG), *electrocardiogram* (ECG), *Galvanic Skin Response* (GSR), and *pupil dilation*, to infer the actual operator's mental workload and opportunely trigger the AA.

Collectively, each strategy has pros and cons. On one hand, the *critical-event approaches* have the great advantage of being absolutely no-invasive and computationally low demanding for the system, since their DFA logic is just based on the occurrence or not of a specific condition. However, their main disadvantage consists in being based on *a priori* models, without taking into account the actual needs of the operator (Parasuraman et al. 1992). On the other hand, *performance measurement* has the advantage of being an "online" method that can potentially reflect the unpredictable changes in the operator's mental state. Nevertheless, this method performs as good as the sensitivity of the measurement technology and algorithms: if the system is not able to detect operator's performance degradation, even though low, it will affect the efficacy of the system in reacting to operator's workload increasing. Also, the additional task for the performance measurement implies a certain intrusiveness of the system that could affect the operator's overall performance (Parasuraman et al. 1992). Performance measurement also occurs "after the fact", i.e. after a passed time period during which the workload became high, so adaptation should be needed to compensate the operator workload increasing (Scerbo et al. 2001). Although the use of both the previous approaches has been successfully employed in several studies, the *neurophysiological measurement-based approach* has several advantages (Byrne and Parasuraman 1996). First, unlike the *critical-event approaches,* neurophysiological measures could be obtained continuously and online. On the other hand, compared with *performance measures*, the neurophysiological ones may be recorded continuously without using overt responses (i.e. additional tasks to measure operator's performance, such as pushing a button and measuring the reaction time) and provide a direct measure of the mental (covert) activities of the operator. In conclusion, neurophysiological measures have higher resolution than performance measures (Di Flumeri et al. 2015a). Finally, the neurophysiological measures can be used not only to trigger the AA, but also to highlight why the AA was needed by the operator, in other words if the operator was under- or overloaded, drowsy, or fatigued. This could potentially offer new perspectives for adaptive intervention to optimize performance by acting on specific aspects of the operator's behavior. The major limits in applying such measures in operational environments are technological and organizational, since the neurophysiological monitoring devices could be perceived as *invasive* by the operators and as *means* to control their performance.

In 1996, Byrne and Parasuraman (1996) assessed that the advantage of applying neurophysiological measures in triggering *AA* was very clear, but the "effective application of psychophysiology in the regulatory role may require years of effort and considerable maturation in technology."

Nowadays, 20 years later, such "effective application" could become reality, thanks to the progresses in *Brain–Computer Interfaces* (BCI) research. In its classical assumption, a BCI is a communication system in which messages or commands that an individual sends to the external world do not pass through the brain's normal output pathways of peripheral nerves and muscles (Wolpaw et al. 2002). More recently, Wolpaw and Wolpaw (2012) defined a BCI as *"a system that measures Central Nervous System (CNS) activity and converts it into artificial output that replaces, restores, enhances, supplements, or improves natural CNS output and thereby changes the ongoing interactions between the CNS and its external or internal environment."* In the BCI community, the possibility of using the BCI systems in different contexts for communication and system control (Aloise et al. 2013)—developing applications in realistic and operational environments—is not just a theory, but something very close to real applications (Blankertz et al. 2010; Zander et al. 2009). In fact, in the classic BCI applications, the user voluntarily modulates its brain activity in order to interact with the system (Vecchiato et al. 2016). In the new BCI concept, called *passive-BCI* (p-BCI), the system itself recognizes the user's spontaneous brain activity related to the considered mental state (e.g. emotional state, workload, attention levels), and uses such information to improve and modulate the interaction between the operator and the system. Systems based on passive-BCI technology can provide objective information about covert aspects of the user's cognitive state, since conventional methods—such as behavioral measures—could only detect such mental states with weak reliability (Zander and Jatzev 2012). The information extracted by the p-BCIs are then employed to improve *Human–Machine Interactions* (HMI) and to achieve potentially novel types of skills. Anyhow, the quantification of mental states using p-BCI technology is far from trivial. In fact, it requires combination of knowledge in different fields (Brouwer et al. 2015), such as neurophysiology (to acquire and manage biosignals), experimental psychology (to find out the right way to assess mental states), machine learning (to develop innovative classification techniques), and human factor (to develop real applications). Mental states monitoring is of particular interest in safety-critical applications where the human factor is often the least controllable factor. In this regard, there are many examples in which p-BCIs could be very useful. For example, p-BCI technology can reveal valuable information about the user's mental state in safety-critical applications, such as driving (Borghini et al. 2012; Maglione et al. 2014; Welke et al. 2009) in the industrial environments (Venthur et al. 2010) or in the context of security surveillance (Marcel and Millan 2007). With respect to driving assistance applications, recent studies demonstrated the utility of p-BCI systems during a driving simulation, for assessing the driving performance and inattentiveness (Schubert and Tangermann 2008), as well as for robustly detecting emergency brakes before braking onset (Welke et al. 2009). In addition, p-BCI systems can potentially be used for real-time cognitive monitoring of the operator's mental workload (Abbass et al. 2014; Arico et al. 2014; Kohlmorgen et al. 2007). These, and other forms of automation, have yielded several benefits in terms of improved capacity, efficiency, and safety. There is evidence that *"although unconstrained handover of automated tasks to operators would increase workload and defeat the purpose of automation, there are situations during which temporary allocation to human control is desirable*" (Parasuraman et al. 1996). During a multitasking flight simulation, they highlighted the superiority of adaptive tasks' allocation (i.e. automation) over static automation—at least for the monitoring aspect of performance. Parasuraman and Rizzo (2008) reported a list of new adaptive solutions that are able—thanks to the technological progresses in the biosensors' domain—to improve the human performance in various environments. Neuroimaging methods and cognitive neuroscience have steadily improved their technical sophistication and breadth of application over the past decades, and there has been a growing interest in using them to examine the neural circuits that support complex tasks representative of perception, cognition, and action (as they occur in operational settings). At the same time, many fields in the biological sciences—including neuroscience—are currently challenged to demonstrate their relevance in practical real-world problems (Parasuraman 2003). Researchers in HF and ergonomics sectors studied the human capabilities and limitations—both cognitive and physical—and used such knowledge to design technologies and work environments to be safer and more usable, efficient, and enjoyable for people who interact with them (Norman and Verganti 2014; Wickens et al. 2012). In today's technology-driven environment, where human capabilities are struggling to keep up with technology offerings, techniques for augmenting the human performance are becoming the critical gap to preclude realizing the full benefits that these technology advances offer. The concept of human performance augmentation is not so recent: the idea was developed during the past decade (Blankertz et al. 2010; Zander et al. 2009), and, at the same time, the concept of Augmented Cognition (*AugCog*) was borne out of the *Defense Advanced Research Projects Agency's* (DARPA), pushing for technologies that enhanced the Warfighter's communication skills and those technologies that involved biosensors for medical applications (Schmorrow 2002). In this context, the mental workload, cognitive control behavior, and training level, due to their strong relationship with the increasing or degrading of the user's performance, are the mental states most studied by the research scientists of the field. The techniques for their investigation must be sensitive to fluctuations in tasks' demands without intruding on primary tasks' performance (O'Donnell and Eggemeier 1986). This level of sensitivity is not obtainable with behavioral (task performance) and subjective measures (e.g. workload perception) alone. In fact, subjective measures, concurrently with the execution of the considered activity, require additional tasks (self-assessment itself). Additionally, it has been widely demonstrated that neurophysiological measurements transcend both behavioral and subjective measures in discriminating cognitive demand fluctuations (Di Flumeri et al. 2015a; Mühl et al. 2014; Wierwille and Eggemeier 1993). Therefore, the neurophysiological measurements become very important, not only as monitoring techniques, but mainly as support tools for the user during his/her operative activities. In fact, as the changes in cognitive activity can be measured in real time, it should also be possible to manipulate the task demand (AA) in order to help the user to keep optimal levels of mental workload under which operating. In other words, the neurophysiological measures could be used to realize p-BCI applications in real operational environments. To overcome the issues and limitations described above, there is the need to focus on the definition and validation of metrics and methodologies capable of addressing such problems and of providing objective and quantitative information about the user's mental status. For such purposes, specific experimental protocols have been designed. The first phase of the researches has been developed and tested in controlled settings (Laboratory), while the consecutive phases have been carried out in realistic environments (e.g. aircrafts, air traffic control room), so involving professional personnel. During the execution of the experimental tasks, the users' neurophysiological signals have been recorded continuously, and both the behavioral data (task performance) and the subjective ones (workload perception) have been gathered with the aim to analyze the considered cognitive phenomena from the different points of view (Triangulation methodology, Bekhet and Zauszniewski 2012). From the biosignals, specific metrics have been defined to investigate the considered cognitive phenomenon. Both standard spectrum EEG analysis and techniques coming from the BCI field (i.e. machine learning approach) have been used to analyze the neurophysiological data. In particular, the *StepWise Linear Discriminant Analysis* (SWLDA) has been chosen since it has been demonstrated to be one of the best outperforming linear classifiers for the mental workload evaluation (Berka et al. 2007; Craven et al. 2006; Krusienski et al. 2006; Rabbi et al. 2009). Before its application, the use of the machine learning approach for the training assessment and workload evaluation in operational environments has been carefully and deeply studied and, as a result, a modified version of the SWLDA, the *automatic stop SWLDA*–asSWLDA, has been developed and patented (Aricò et al. 2015a). The asSWLDA has been able to solve important key issues in using neurophysiological measurements in operational environments: (i) over-time reliability of the measure (no recalibration of the algorithm is necessary within a month), (ii) automatic features selection to avoid both the *underfitting* and *overfitting* problems (no empiric and manual selection of the analysis parameters), and (iii) it provided higher accuracy (resolution) in comparison with the standard (i.e. subjective) workload measures.

## References

Abbass HA, Tang J, Amin R, Ellejmi M, Kirby S (2014) Augmented cognition using real-time EEG-based adaptive strategies for air traffic control. Proc Hum Factors Ergon Soc Annu Meet 58:230–234. doi:10.1177/1541931214581048

Aberg L, Rimmö PA (1998) Dimensions of aberrant driver behaviour. Ergonomics 41:39–56. doi:10.1080/001401398187314

Aloise F, Aricò P, Schettini F, Salinari S, Mattia D, Cincotti F (2013) Asynchronous gaze-independent event-related potential-based brain–computer interface. Artif Intell Med 59:61–69. doi:10.1016/j.artmed.2013.07.006. Special Issue: Brain-computer interfacing

Aricò P, Borghini G, Di Flumeri G, Babiloni F (2015) Metodo per stimare uno stato mentale, in particolare un carico di lavoro, e relativo apparato (A Method for the estimation of mental state, in particular of the mental workload and its device). P1108IT00

Arico P, Borghini G, Graziani I, Taya F, Sun Y, Bezerianos A, Thakor NV, Cincotti F, Babiloni F (2014) Towards a multimodal bioelectrical framework for the online mental workload evaluation. Presented at the 2014 36th annual international conference of the IEEE Engineering in Medicine and Biology Society (EMBC), pp. 3001–3004. doi:10.1109/EMBC.2014.6944254

Bekhet AK, Zauszniewski JA (2012) Methodological triangulation: an approach to understanding data. Nurse Res. 20:40–43. doi:10.7748/nr2012.11.20.2.40.c9442

Berka C, Levendowski DJ, Lumicao MN, Yau A, Davis G, Zivkovic VT, Olmstead RE, Tremoulet PD, Craven PL (2007) EEG correlates of task engagement and mental workload in vigilance, learning, and memory tasks. Aviat Space Environ Med 78:B231–B244

Blankertz B, Tangermann M, Vidaurre C, Fazli S, Sannelli C, Haufe S, Maeder C, Ramsey L, Sturm I, Curio G, Müller K-R (2010) The Berlin Brain–Computer Interface: non-medical uses of BCI technology. Front Neurosci 4. doi:10.3389/fnins.2010.00198

Borghini G, Vecchiato G, Toppi J, Astolfi L, Maglione A, Isabella R, Caltagirone C, Kong W, Wei D, Zhou Z, Polidori L, Vitiello S, Babiloni F (2012) Assessment of mental fatigue during car driving by using high resolution EEG activity and neurophysiologic indices. Annual international conference of the IEEE Engineering in Medicine and Biology Society (EMBC) 2012, pp 6442–6445. doi:10.1109/EMBC.2012.6347469

Brouwer A-M, Zander TO, van Erp JBF, Korteling JE, Bronkhorst AW (2015) Using neurophysiological signals that reflect cognitive or affective state: six recommendations to avoid common pitfalls. Front Neurosci 9:136. doi:10.3389/fnins.2015.00136

Byrne EA, Parasuraman R (1996) Psychophysiology and adaptive automation. Biol Psychol 42:249–268. doi:10.1016/0301-0511(95)05161-9. Psychophysiology of Workload

Calabrese EJ (2008) Neuroscience and hormesis: overview and general findings. Crit Rev Toxicol 38:249–252. doi:10.1080/10408440801981957

Craven PL, Belov N, Tremoulet P, Thomas M, Berka C, Levendowski D, Davis G (2006) Cognitive workload gauge develpment: comparison of real-time classification methods. Augment Cogn. Past Present Future

Decades NRC (US) Military communities, IM for EN, CR in the NT (2008) Emerging areas of cognitive Neuroscience and Neurotechnologies

Di Flumeri G, Borghini G, Aricò P, Colosimo A, Pozzi S, Bonelli S, Golfetti A, Kong W, Babiloni F (2015) On the use of cognitive neurometric indexes in aeronautic and air traffic management environments. In: Blankertz B, Jacucci G, Gamberini L, Spagnolli A, Freeman J (eds) Symbiotic interaction. Springer International Publishing, Cham, pp 45–56

Feyer A, Williamson AM (2011) Human factors in accident modelling. Encycl Occup, Health Saf

Fuchs S, Hale KS, Stanney KM, Juhnke J, Schmorrow DD (2007) Enhancing mitigation in augmented cognition. J Cogn Eng Decis Mak 1:309–326. doi:10.1518/155534307X255645

Giraudet L, Imbert J-P, Bérenger M, Tremblay S, Causse M (2015) The neuroergonomic evaluation of human machine interface design in air traffic control using behavioral and EGG/ERP measures. Behav Brain Res 294:246–253. doi:10.1016/j.bbr.2015.07.041

Helmreich RL (2000) On error management: lessons from aviation. BMJ 320:781–785

Hilburn B, Jorna PG, Byrne EA, Parasuraman R (1997) The effect of adaptive air traffic control (ATC) decision aiding on con-troller mental workload. In: Mouloua M, Koonce J (eds) Human-automation interaction: research and practice. Erlbaum, Mahwah, NJ, pp 84–91

Hoc JM (2000) From human-machine interaction to human-machine cooperation. Ergonomics 43:833–843. doi:10.1080/001401300409044

Inagaki T (2003) Adaptive automation: sharing and trading of control. Handbook cognitive task design. Erlbaum, Mahwah. pp 147–169

Kaber DB, Riley JM, Tan K-W, Endsley MR (2001) On the design of adaptive automation for complex systems. Int J Cogn Ergon 5:37–57. doi:10.1207/S15327566IJCE0501_3

Kirsh D (2000) A few thoughts on cognitive overload. Intellectica 30

Kirwan B (1998) Human error identification techniques for risk assessment of high risk systems–Part 1: Review and evaluation of techniques. Appl Ergon 29:157–177
Kohlmorgen J, Dornhege G, Braun M, Blankertz B, Müller K-R, Curio G, Hagemann K, Bruns A, Schrauf M, Kincses W (2007) Improving human performance in a real operating environment through real-time mental workload detection [WWW Document]
Krusienski DJ, Sellers EW, Cabestaing F, Bayoudh S, McFarland DJ, Vaughan TM, Wolpaw JR (2006) A comparison of classification techniques for the P300 speller. J Neural Eng 3:299–305. doi:10.1088/1741-2560/3/4/007
Lawton R, Ward NJ (2005) A systems analysis of the Ladbroke Grove rail crash. Accid Anal Prev 37:235–244. doi:10.1016/j.aap.2004.08.001
Maglione A, Borghini G, Aricò P, Borgia F, Graziani I, Colosimo A, Kong W, Vecchiato G, Babiloni F (2014). Evaluation of the workload and drowsiness during car driving by using high resolution EEG activity and neurophysiologic indices. Annual international conference of the IEEE Engineering in Medicine and Biology Society (EMBC), 2014, pp 6238–6241. doi:10.1109/EMBC.2014.6945054
Marcel S, Millan JDR (2007) Person authentication using brainwaves (EEG) and maximum a posteriori model adaptation. IEEE Trans Pattern Anal Mach Intell 29:743–752. doi:10.1109/TPAMI.2007.1012
Mühl C, Jeunet C, Lotte F (2014) EEG-based workload estimation across affective contexts. Neuroprosthetics 8:114. doi:10.3389/fnins.2014.00114
Nelson WR, Haney LN, Ostrom LT, Richards RE (1998) Structured methods for identifying and correcting potential human errors in space operations. Acta Astronaut 43:211–222
Norman DA, Verganti R (2014) Incremental and radical innovation: Design research vs. technology and meaning change. Design issues 30(1):78–96
O'Donnell RD, Eggemeier (1986) Workload assessment methodology. Handbook of perception and human performance. John Wiley and Sons, Inc
Parsons MW, Harrington DL, Rao SM (2005) Distinct neural systems underlie learning visuomotor and spatial representations of motor skills. Hum Brain Mapp 24:229–247. doi:10.1002/hbm.20084
Parasuraman R (2003) Neuroergonomics: research and practice. Theor Issues Ergon Sci 4:5–20. doi:10.1080/14639220210199753
Parasuraman R, Rizzo M (2008) Neuroergonomics: the brain at work, 1, edition edn. Oxford University Press, New York
Parasuraman R, Bahri T, Deaton JE, Morrison JG, Barnes M (1992) Theory and design of adaptive automation in aviation systems
Parasuraman R, Mouloua M, Molloy R (1994) Monitoring automation failures in human-machine systems. In: Mouloua M, Parasuraman R (eds) Human performance in automated systems: current research and trends. LEA, Hillsdale, NJ, pp 45–49
Parasuraman R, Mouloua M, Molloy R (1996) Effects of adaptive task allocation on monitoring of automated systems. Hum Factors 38:665–679
Rabbi AF, Ivanca K, Putnam AV, Musa A, Thaden CB, Fazel-Rezai R (2009) Human performance evaluation based on EEG signal analysis: a prospective review. Annual international conference of the IEEE Engineering in Medicine and Biology Society (EMBC), 2009, pp 1879–1882. doi:10.1109/IEMBS.2009.5333877
Rankin W, Hibit R, Allen J, Sargent R (2000) Development and evaluation of the Maintenance Error Decision Aid (MEDA) process. Int J Ind Ergon 26:261–276. doi:10.1016/S0169-8141(99)00070-0
Rasmussen J (1987) The definition of human error and a taxonomy for technical system design. New technology and human error, pp 23–30
Rasmussen J (1982) Human errors. A taxonomy for describing human malfunction in industrial installations. J Occup Accid 4:311–333. doi:10.1016/0376-6349(82)90041-4
Reason J (2008) The human contribution: Unsafe acts, accidents and heroic recoveries, New edition. CRC Press, Farnham, England ; Burlington, VT
Reason J (2016) Managing the risks of organizational accidents. Routledge

Rouse WB (1976) Adaptive allocation of decision making responsibility between supervisor and computer. In: Sheridan TB, Johannsen G (eds) Monitoring behavior and supervisory control, NATO conference series. Springer, US, pp 295–306. doi:10.1007/978-1-4684-2523-9_24
Rumar K (1990) The basic driver error: late detection. Ergonomics 33:1281–1290. doi:10.1080/00140139008925332
Salmon P, Regan M, Johnston I (2005) Human error and road transport: phase one—literature review. Monash University Accident Research Centre
Scerbo MW, Freeman FG, Mikulka PJ (2003) A brain-based system for adaptive automation. Theor Issues Ergon Sci 4:200–219. doi:10.1080/1463922021000020891
Scerbo MW, Freeman FG, Mikulka PJ, Parasuraman R, Nocero FD, III LJP (2001) The efficacy of psychophysiological measures for implementing adaptive technology
Schmorrow DD (2002) DARPA's augmented cognition program-tomorrow's human computer interaction from vision to reality: building cognitively aware computational systems. In: Presented at the proceedings of the 2002 IEEE 7th conference on human factors and power plants, 2002, pp 7-1–7-4. doi:10.1109/HFPP.2002.1042859
Schubert R, Tangermann M (2008) "Parieto-occipital alpha power indexes distraction during simulated car driving," abstracts of the 14th World congress of psychophysiology. Int J Psychophysiol 69. doi:10.1016/j.ijpsycho.2008.05.033
Sexton JB, Thomas EJ, Helmreich RL (2000) Error, stress, and teamwork in medicine and aviation: cross sectional surveys. BMJ 320:745–749
Shappel SA, Wiegmann DA (2000) The human factors analysis and classification system-HFACS (No. DOT/FAA/AM-00/7). Federal Aviation Administration, Washington, DC
Sheridan TB (2002) Humans and automation: system design and research issues. John Wiley & Sons Inc, New York, NY, USA
Sheridan TB (1992) Telerobotics, automation, and human supervisory control. MIT Press, Cambridge
Sheridan TB, Verplank WL (1978) Human and computer control of undersea teleoperators
Shorrock ST, Kirwan B (2002) Development and application of a human error identification tool for air traffic control. Appl Ergon 33:319–336
Stanton NA, Harris D, Salmon PM, Demagalski JM, Marshall A, Young MS, Dekker SWA, Waldmann T (2006) Predicting design induced pilot error using HET (human error template)—A new formal human error identification method for flight decks [WWW Document]. https://dspace.lib.cranfield.ac.uk/handle/1826/1158. Accessed 25 Nov 2015
Vecchiato G, Borghini G, Aricò P, Graziani I, Maglione AG, Cherubino P, Babiloni F (2016) Investigation of the effect of EEG-BCI on the simultaneous execution of flight simulation and attentional tasks. Med Biol Eng Comput 54:1503–1513. doi:10.1007/s11517-015-1420-6
Venthur B, Blankertz B, Gugler MF, Curio G (2010) Novel applications of BCI technology: psychophysiological optimization of working conditions in industry. In: Presented at the 2010 IEEE International Conference on Systems Man and Cybernetics (SMC), IEEE, pp 417–421. doi:10.1109/ICSMC.2010.5641772
Warm JS, Parasuraman R, Matthews G (2008) Vigilance requires hard mental work and is stressful. Hum Factors 50:433–441
Welke S, Jurgensohn T, Roetting M (2009) Single-trial detection of cognitive processes for increasing traffic safety. In: Proceedings of the 21st (Esv) international technical conference on the enhanced safety of vehicles. Held June 2009, Stuttgart, Germany
Wickens CD (1992) Engineering psychology and human performance, 2nd edn. HarperCollins Publishers, New York, NY, US
Wickens CD, Hollands JG, Banbury S, Parasuraman R (2012) Engineering psychology and human performance, 4th edn. Psychology Press, Boston
Wierwille WW, Eggemeier FT (1993) Recommendations for mental workload measurement in a test and evaluation environment. Hum Factors J Hum Factors Ergon Soc 35:263–281. doi:10.1177/001872089303500205
Wolpaw J, Wolpaw EW (2012) Brain-Computer Interfaces: principles and practice. Oxford University Press, Oxford

Wolpaw JR, Birbaumer N, McFarland DJ, Pfurtscheller G, Vaughan TM (2002) Brain-Computer Interfaces for communication and control. Clin Neurophysiol Off J Int Fed Clin Neurophysiol 113:767–791

Zander TO, Jatzev S (2012) Context-aware brain-computer interfaces: exploring the information space of user, technical system and environment. J Neural Eng 9:016003. doi:10.1088/1741-2560/9/1/016003

Zander TO, Kothe C, Welke S, Roetting M (2009) Utilizing secondary input from passive Brain-Computer Interfaces for enhancing human-machine interaction. In: Schmorrow DD, Estabrooke IV, Grootjen M (eds) Foundations of augmented cognition. Neuroergonomics and operational neuroscience. Springer, Berlin, Heidelberg, pp 759–771

# Chapter 2
# Preliminary Concepts

In this chapter of the book, we illustrate the brain structures, properties of the central nervous system, and generation of the electroencephalogram (EEG) signal. In particular, from the organization of the central nervous system and its peripheral organs, we describe the neuroanatomy of the brain cortex, and how neurons generate the electric potentials and create the synapses to consolidate the memory and send messages to the different parts of the brain and of the body. In fact, neurons are the basic units of the nervous system and are specialized to generate electrical signals, which are then used to encode and convey information: these signals are expressed by alterations of the resting membrane potential. The synchronization of neurons activity within specific frequency ranges define the characteristic rhythms (or bands) of the EEG. The scientific literature widely demonstrated how such EEG rhythms are correlated to specific cognitive processes, and brain lobes to specific functions and behavior, such as auditory rather than visual stimuli processing, long-term memory, and language. Therefore, the analysis of the EEG rhythms across the brain and how they communicate allow to investigate the different user's cognitive functions and mental states.

## 2.1 Nervous System

*Gross Anatomy of the Brain*

The nervous system essentially exhibits a bilateral symmetry, with structural features and pathways located on one side of the midline (Noback et al. 2005). It is subdivided anatomically into the *Central Nervous System* (CNS) and the *Peripheral Nervous System* (*PNS*), while functionally into the *Somatic Nervous System* (SNS) and the *Autonomic* (visceral) *Nervous System* (ANS). The CNS comprises the brain and the spinal cord. The brain is encapsulated within the skull, while the spinal cord is located at the center of the vertebral column. The PNS consists of the



G. Borghini et al., *Industrial Neuroscience in Aviation*,
Biosystems & Biorobotics 18, DOI 10.1007/978-3-319-58598-7_2

nerves emerging from the brain (called *cranial nerves*) and from the spinal cord (called *spinal nerves*). The peripheral nerves convey neural messages from (1) the sense organs and sensory receptors in the organism inward to the CNS, and from (2) the CNS outward to the muscles and to the glands of the body. The SNS consists of those neural structures of the CNS and PNS responsible for (1) conveying and processing conscious and unconscious sensory (*afferent*) information, vision, pain, touch, unconscious muscle sense from the head, body wall, and extremities to the CNS; (2) motor (*efferent*) control of the voluntary (striated) muscles. The ANS is composed of the neural structures responsible for (1) conveying and processing sensory input from the visceral organs (e.g. digestive system and cardiovascular system); (2) motor control of the involuntary (smooth) and cardiac musculature, and of the glands of the viscera. The sensory signals originating in the sensory receptors are transmitted through the nervous system along sensory pathways, e.g. pain and temperature pathways and visual pathways. These signals may reach consciousness or may be utilized at unconscious levels. The neural messages for the motor activity are conveyed through the nervous system to the muscles and the glands, along motor pathways. Both the sensory pathways (ascending) and the motor pathways (descending) include processing centers for each pathway (e.g. ganglia, nuclei, laminae, cortices), located at different anatomic levels of the spinal cord and of the brain. The processing centers are the computers of the complex high-speed systems within the brain. Differences in the basic sequence are present in some ascending systems. In a general way, the motor systems are organized so as to receive stimuli from the sensory systems, at all levels of the spinal cord and brain, and to convey messages via motor pathways to neuromuscular and neuroglandular, endings at muscle and gland cells in the head, body, and extremities. The motor pathways comprise sequences of processing centers, and their fibers conveying neural influences to other processing centers within the CNS, and the final linkages extending from the CNS via motor nerves of the PNS to muscles and glands. The CNS comprises *gray matter* and *white matter* (Fig. 2.2). The gray matter consists of neuronal cell bodies, dendrites, axon terminals, synapses, and glial cells, and is highly vascular. The white matter—whose color is imparted by the myelin—consists instead of bundles of axons (many of which are myelinated and oligodendrocytes); it lacks of neuronal cell bodies and is less vascular than the gray matter. Groupings of neuronal cell bodies within the gray matter are variously known as nucleus, ganglion, lamina, body, cortex, center, formation, and horn. A cortex is a layer of gray matter on the surface of the brain. Two major cortices are recognized: *cerebral* and *cerebellar* cortices. The superior and inferior colliculi of the midbrain and the hippocampal formation also form cortex-like structures.

The *cerebrum* includes the paired cerebral hemispheres, a small median segment (derived from the telencephalon), and the diencephalon. The *cerebral hemispheres* consist of the cerebral cortex (gray matter) that underlies the white matter, the corpus striatum and corpus callosum, the anterior commissure, the hippocampal formation, and the amygdala (Fig. 2.1). The brain hemispheres are marked on the surface by slit-like incisures called *sulci*, and the raised ridge between the two sulci

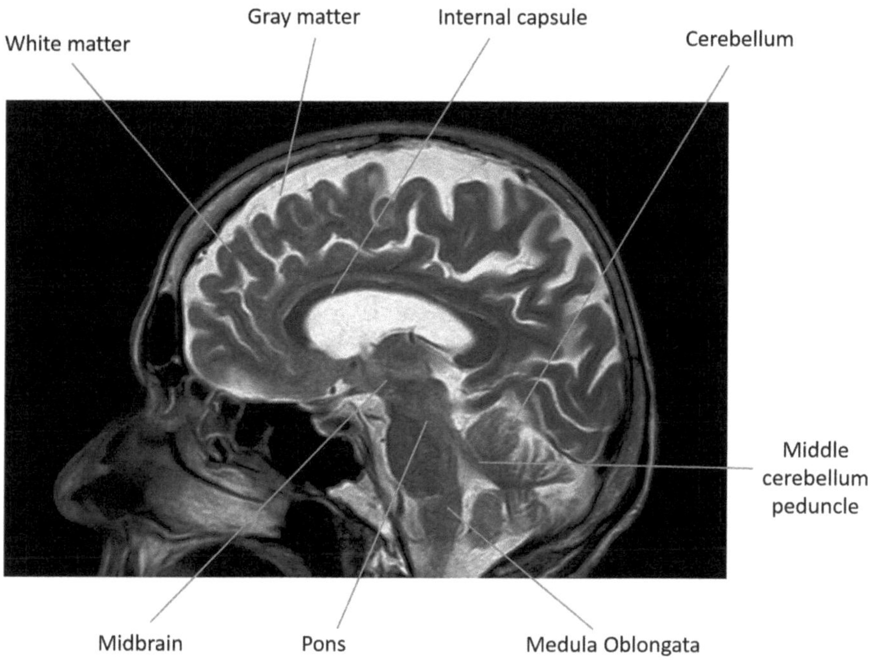


**Fig. 2.1.** A sagittal brain section in which some of the main structures of the brains are labeled. In particular, gray matter and white matter, cerebellum, midbrain, medulla oblongata, pons, middle cerebellar peduncle, and internal capsule

is called *gyrus*. The cortex lining a sulcus is considered part of the adjacent gyrus, while the hemispheres are separated from one another in the midline by the longitudinal *fissure* (particularly deep and constant sulcus). Each hemisphere is conventionally divided into six *lobes*: frontal, parietal, occipital, temporal, central (insula), and limbic (Fig. 2.2). The portion of the frontal, parietal, and temporal lobes that overlies the insula is called *operculum*. The lobes are delineated one from another by several major sulci, including the lateral sulcus of Sylvius, the central one of Rolando, the cingulate, and the parieto-occipital sulci. The *lateral sulcus* is a deep furrow that outspreads posteriorly from the basal surface of the brain along the lateral surface of the hemisphere, terminating usually as an upward curve within the inferior part of the parietal lobe. The *central sulcus* of Rolando, instead, extends obliquely from the region of the lateral sulcus across the dorsolateral cerebral surface and, for a short distance, onto the medial surface. The *cingulate sulcus* is a curved cleft on the medial surface, lengthening parallel to the curvature of the corpus callosum. The *parieto-occipital sulcus* is a deep cleft on the medial surface located between the central sulcus and the occipital pole (Fig. 2.3).

The lobes on the lateral cerebral surface are situated on the following boundaries: (1) the frontal lobe is in front of the central sulcus and above the lateral sulcus; (2) the occipital lobe is located behind an imaginary line parallel to the

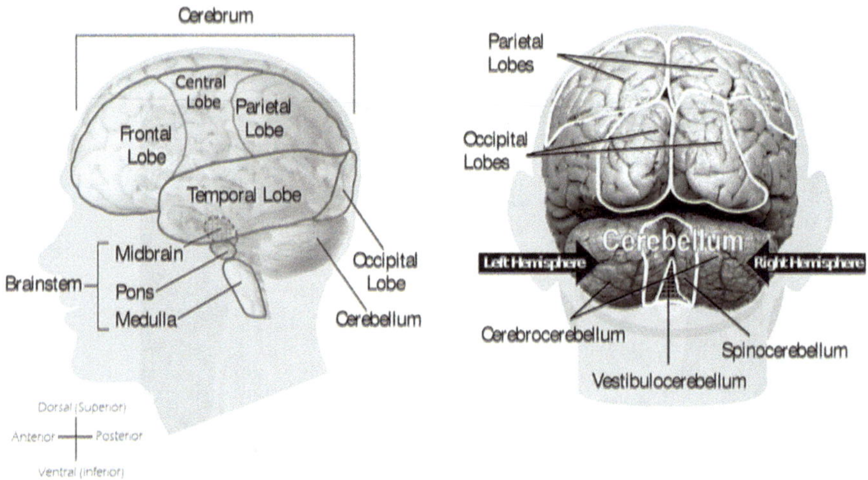


**Fig. 2.2** Brain lobes and hemispheres

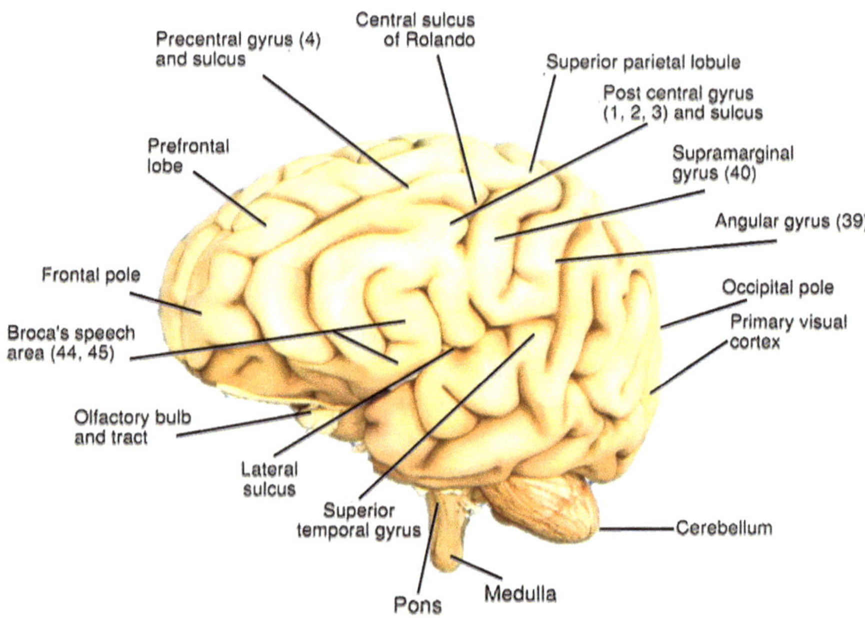


**Fig. 2.3.** Medial surface of the cerebral hemisphere. The limbic lobe consists of the cingulate gyrus, the isthmus and the parahippocampal gyrus. The amygdala is located within the uncus. The hippocampal formation (hippocampus and dentate gyrus) is located in the floor of the temporal horn of the lateral ventricle

parieto-occipital sulcus, which is situated on the medial surface; (3) the parietal lobe is located at the rear of the central sulcus, anteriorly to the imaginary parieto-occipital line and above the lateral sulcus; it runs out toward the occipital pole before taking an upward curve; (4) the temporal lobe is positioned below the lateral sulcus, and before the imaginary parieto-occipital line; and (5) the central lobe is placed at the bottom (medial surface) of the lateral sulcus of Sylvius, which is actually a deep fossa (depression). It is visible only when the temporal and frontal lobes are reflected away from the lateral sulcus (Fig. 2.3). Instead, the lobes on the medial cerebral surface are positioned as follows: (1) the frontal lobe is located rostral to a line formed by the central sulcus; (2) the parietal lobe is situated between the central sulcus and the parieto-occipital sulcus; (3) the temporal lobe is located lateral to the parahippocampal gyrus; (4) the occipital lobe is posterior to the parieto-occipital sulcus; and (5) the limbic lobe is a synthetic one, formed by parts of the frontal, parietal, and temporal lobes. It is located in the center of the curved line formed by the cingulate sulcus and the collateral sulcus (the latter is located lateral to the parahippocampal gyrus). The limbic lobe is the ring (limbus) of gyri bordered by this line; it includes the subcallosal area, the cingulate gyrus, the parahippocampal gyrus, the hippocampus, the dentate gyrus, and the uncus.

*Basic Unit of the Nervous System: Neurons*

Some 100–200 billion ($[1–2] \times 10^{11}$) *neurons*, as well as many more glial cells, are integrated into the structural and functional structures of the brain (Noback et al. 2005). They exhibit a wide diversity of shapes and sizes. The neuron is the basic unit of the nervous system and is composed of four regions, structurally defined as follows: *a cell body* (soma) that emits a single nerve process called an *axon*, which ends with the *presynaptic terminals*, and a variable number of branching processes called *dendrites* (Fig. 2.4). Each axon, including its collateral branches, usually terminates as an arbor of fine fibers; each fiber ends as an enlargement called *bouton*, which is part of a synaptic junction. At the other end of the neuron, there is a three-dimensional *dendritic field*, formed by the branching of the dendrites. The cell body is the genomic and metabolic center of the neuron. Dendrites are the main recipients of the neural signals for communication between neurons and contain critical processing complexes. The axon is the canal apt to conduct messages (action potentials) to the presynaptic terminals, where each neuron is in synaptic contact with other neurons and, thus, is part of the network that constitutes the nervous system. A neuron is designed to react to the stimuli, to rapidly transmit the resulting excitation to the other portions of the nerve cell, and to influence other neurons, muscle cells, and glandular cells. Neurons are so specialized that most are incapable of reproducing themselves, and they lose viability if denied an oxygen supply for more than a few minutes. Dendrites contain the same cytoplasmic organelles (e.g. Nissl bodies and mitochondria) as the cell body of which they are true extensions. The axon is specialized in the transmission of coded information as all-or-none action potentials. The axon arises from the *axon hillock* of the cell body,

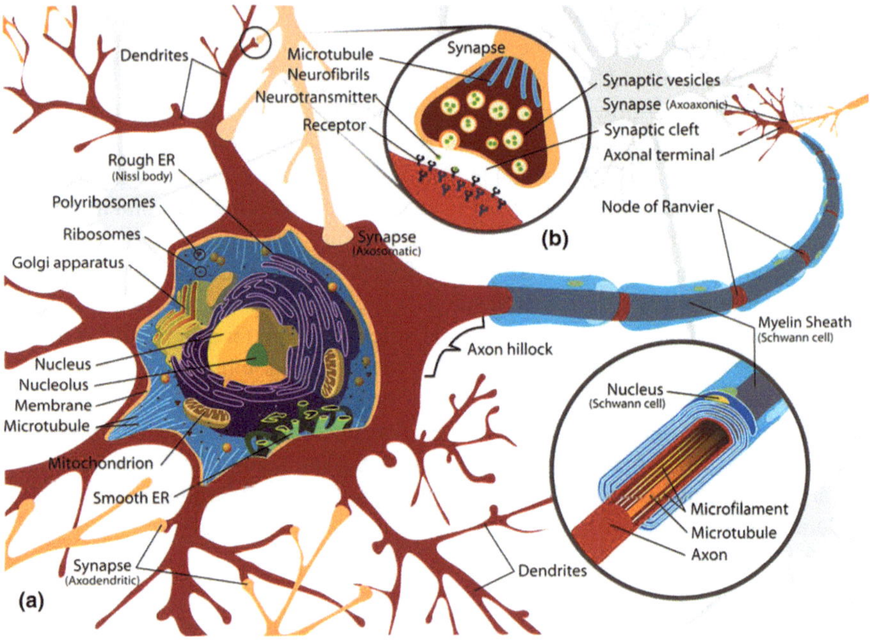


**Fig. 2.4.** **a** Diagram of a neuron located wholly within the central nervous system. The **b** represents a synapse between two neurons. The myelin sheath of neuron **a** is entirely the product of a glial cell, and the one of neurons is produced by a glial cell inside the central nervous system and by a Schwann (neurolemma) cell in the PNS

at a site called the *initial segment,* and extends for a distance of less than 1 (mm) to as much as 1 (m) before arborizing into *terminal branches* (Fig. 2.4). The axon hillock, initial segment, and the axon lack Nissl bodies. The branches of an axon can have, potentially, two types of *bouton*. Each branch ends as a *terminal bouton* that forms a synapse with the dendrite, cell body, or axon of another neuron. In addition, along some branches, there are thickenings called *boutons en passage,* which form synapses with another neuron or smooth muscle fiber. The dendrites of many neurons are studded with tiny protuberances called *spines* (e.g. pyramidal neurons of the cerebral cortex): these dendritic spines increase the surface area of the membrane of the receptive segment of the neuron. Located on them, there are over 90% of all the excitatory synapses present in the central nervous system (CNS). Because of their widespread occurrence on neurons of the cortical areas of the cerebrum, they are thought to be involved in learning and memory (see Para 3.3).

The synapse is the contact point between the neurons. A submicroscopic space, the synaptic cleft, which is about 200 (Å), exists between the bouton of one neuron and the cell body of another neuron (*axosomatic synapse*), between a bouton and a dendrite (*axodendritic synapse*), and between a bouton and an axon (*axoaxonic synapse*). In addition, also *dendrodendritic synapses* (between two dendrites).

The axon of one neuron might terminate in just a few synapses or up to many thousands of synapses. The dendrite–cell body complex might receive synaptic contacts from many different neurons (up to over 15,000 synapses). The termination of a nerve fiber in a muscle cell (neuromuscular junction) or a glandular cell (neuroglandular junction) is basically similar to the synapse between two neurons. The synapse of each axon terminal of a motoneuron on a voluntary muscle cell is called a *motor end plate* (Fig. 2.4). The cell membrane of the axon at the synapse is the *presynaptic membrane*, and the cell membrane of dendrite–cell body complex, muscle, or glandular cell is the *postsynaptic membrane*. The *subsynaptic membrane* is that region of the postsynaptic membrane that is juxtaposed against the presynaptic membrane at the synapse. A concentration of mitochondria and *presynaptic vesicles* is present in the cytoplasm of the bouton; none is present in the cytoplasm adjacent to the subsynaptic membrane. Most neurons contain at least two distinct types of vesicle: small vesicles 50 (nm) in diameter and large vesicles from 70 to 200 (nm) in diameter.

### 2.1.1 Basic Neurophysiology

Every neuron is said to have "in small-scale, the integrative capacity of the entire nervous system." In fact neurons can transform information and transmit it to other neurons. In most, the dendrite–cell body unit is specialized as a receptor and integrator of synaptic input from other neurons, and the axon is specialized to convey coded information from the dendrite–cell body unit to the synaptic junctions, where transformation functions take place with other neurons or effectors (muscles and glands). To serve these tasks, the neuron is thus organized into a receptive segment (dendrites and cell body), a conductile segment (axon), and an effector segment (synapse) (Fig. 2.5). Neurons are specialized to generate electrical signals, which are then used to encode and convey information: these signals are expressed by alterations in the resting membrane potential. Voltage changes that are restricted to the sites where neurons are stimulated—or that are close to them—are called *graded potentials*. These ones can lead to the production of *action potentials* (*nerve impulses or spikes*), which transmit information for substantial distances along an axon. Two forms of graded potential are *generator* (*receptor potentials*) and *synaptic potentials*. Generator potentials are evoked by sensory stimuli from the environment (both inside and outside the body). Information that passes from one neuron to another at synapses produces *synaptic potentials* in the postsynaptic neuron. The activity of either generator or synaptic potentials can elicit action potentials, which, in turn, produce synaptic potentials in the next neuron. Synaptic potentials elicited in effectors (skeletal muscle and glands) at synapses can result in the contraction of the muscle or emission of secretory product from a gland.

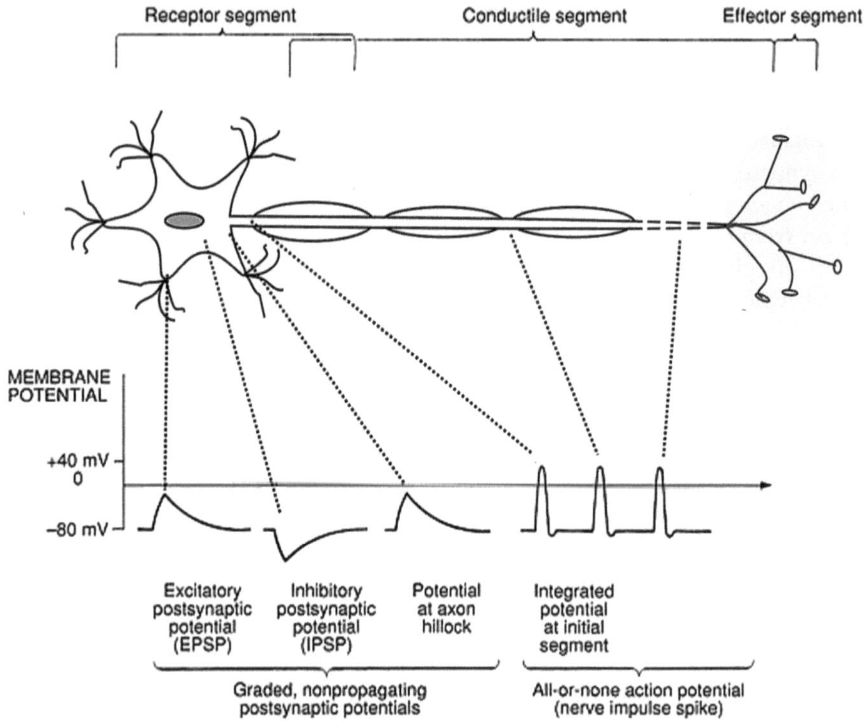


**Fig. 2.5.** On the surface of the dendrites and cell body are excitatory and inhibitory synapses, which, when stimulated, produce local, graded, non-propagating potentials. These are exhibited as an excitatory or depolarizing postsynaptic potential (EPSP) and as an inhibitory or hyperpolarizing postsynaptic potential (IPSP). These local potentials are summated at the axon hillock and, if adequate, could trigger an integrated potential at the initial segment and an "all-or-none" action potential, which is conducted along the axon to the motor end plate

### *Resting Potential of the Neuron*

The resting neuron is a charged cell that is not conducting a nerve impulse. The plasma membrane, which acts as a thin boundary between the extracellular (interstitial) fluid that is located outside the neuron and the intracellular fluid (neuroplasm) that is instead inside the neuron, is critical for maintaining this charged state or resting potential. The electric charge across the plasma membrane results form a thin film of positive and negative ions, unequally distributed across the membrane. These are sodium ($Na^+$) and chloride ($Cl^-$) ions (which are in higher concentration in the interstitial fluid), and potassium ($K^+$) and protein (organic) ions that are in higher concentration in the neuroplasm. A tendency exists for the $Na^+$, $K^+$, and $Cl^-$ ions to diffuse across the membrane from regions of high to low concentration (along concentration gradients), through $Na^+$, $K^+$, and $Cl^-$ channels, respectively. The passage of ions across the membrane is known as *conductance*. Thus, the semipermeable plasma membrane is selectively permeable through non-gated open channels to $Na^+$, $K^+$, and $Cl^-$ ions and impermeable to large protein ions. These channels, which are always open, are important in determining the resting potential. The ionic concentrations on either side of the membrane are produced and maintained by a system of membrane pumps called "the *sodium–potassium pump*", *which* requires metabolic energy released by adenosine triphosphate (ATP). The sodium–potassium exchange pump is an integral membrane protein that utilizes ATP as an energy source for its role in *active transport*. This transport is an energy-dependent process in which the movement of $Na^+$ and $K^+$ ions is "uphill" against a concentration gradient. The activity of the pump results in the passage of three $Na^+$ ions out of and two $K^+$ ions into the neuron. This causes the restoration of a concentration of $K^+$, 30 or more times higher within the neuroplasm than in the interstitial fluid, and in a concentration of $Na^+$ that is 10 times and $Cl^-$ that is 14 times higher in the interstitial fluid than in the neuroplasm. Most neurons do not have a $Cl^-$ pump; hence, $Cl^-$ ions diffuse passively across the membrane. These are the ionic concentrations responsible for establishing an electric potential across the membrane. The transmembrane potential, known as the *resting potential*, is about −70 to −80 (mV) (millivolts) inside the neuron. The resting potential is in a steady state (*dynamic equilibrium*) requiring metabolic energy to maintain the ionic gradients across the membrane. When the neuron is "at rest," its membrane potential is the result of a balance (involving $Na^+$ and $K^+$ ions) between the active fluxes (movements) of ions metabolically driven by *pumps* and the passive fluxes caused by *diffusion*. The active fluxes result from the pump extruding three $Na^+$ ions for every two $K^+$ ions it brings into the neuron. The passive fluxes of ions take place through non-gated channels. The outward flux of positive charges by the pump tends to hyperpolarize the membrane. The greater the hyperpolarization, the greater the inward electrochemical force driving $Na^+$ into the neuron, and the smaller the force driving $K^+$ out. The steady state for the neuron is attained when the resting potential is reached at the point when the net passive inward current (movement of electrical charge) through the ion channels exactly counterbalances the active outward current driven by the pump. The steady state is not basically the result of *passive diffusion*, which is the diffusion of a solute down a concentration gradient without the expenditure of energy.

*Excitability of the Neuron*

Excitability is a property that enables a neuron to respond to a stimulus and to transmit information in the form of electrical signals. The flow of information within a neuron and between neurons is conveyed by both electrical and chemical signals. The electrical signals, called *graded potentials* and *action potentials*, are all produced by temporary changes in the current flow into and out of the neuron—changes that are actually deviations from the normal value of the resting membrane potential. Ion channels within the plasma membrane control instead the inward and outward current flow. The channels have three features: (1) they conduct ions across the plasma membrane at rapid rates up to 100,000,000 ions per second; (2) they can recognize specific ions and be selective as to which can pass through; (3) they can selectively open and close, in response to specific electrical, chemical, and mechanical stimuli.

Each neuron is presumed to have over 20 different types of channel with thousands of copies of each channel. The flux (movement of ions) through the ion channels is passive, requiring no expenditure of metabolic energy. The flux direction is determined by the electrochemical driving force across the plasma membrane, and the primary role of the ion channels in the neurons is to mediate rapid signaling. These channels, called *gated channels*, have a molecular “cap” or *gate*, which briefly opens to permit anion species to pass. Gated channels open when a neurotransmitter binds to them; *voltage-gated channels* open and close in response to changes in membrane potential; *modality-gated channels* are activated by specific modalities (e.g. touch, pressure, or stretch). Gating is the process by which a channel is opened or closed during activity. Each channel consists of several plasma membrane-spanning polypeptide subunits (proteins) arranged around a central pore. Each of these classes of channel belongs to a different gene family. Each member of a family shares common structural and biochemical features, which presumably have evolved from a common ancestral gene of that family. The channels of the voltage-gated gene family are selective for $Na^+$, $K^+$, and $Ca^{2+}$ ions. The channels for the transmitter-gated channels respond to acetylcholine, gamma amino butyric acid (GABA), and glycine. Most gated channels are closed with the membrane at rest: they open when activated, following the binding of a ligand (ligand gating), a change in the membrane potential (voltage gating), or the stretch of the membrane (modality gating). In the transmitter-gated channel, the transmitter binds to a specific site on the external face of a channel that activates it to open briefly. The energy to open the channels is derived from three sources: (1) from the binding of the transmitter to the receptor protein in the ligand-gated channels; (2) from the changes in the membrane voltage within the voltage-gated channels; (3) presumably, from the mechanical forces resulting from cytoskeletal interaction at the modality-gated channels. There are two types of membranes’ response: (1) *hyperpolarization* or (2) *depolarization*. During the *hyperpolarization*, the membrane becomes more negative on the inside with respect to its outside [i.e. could go from −70 (mV) to −80 (mV)]. During *depolarization,* the membrane becomes less negative inside with respect to its outside and it might even reverse polarity with its inside—becoming positive with respect to the outside. This is still called *depolarization* because the membrane potential becomes less negative than the resting potential [e.g. from −70 (mV) to 0 to −40 (mV)].

### 2.1.2 EEG Generation

The *EEG* comes from the summation of synchronously postsynaptic potentials. The contribution to the electric field of neurons acting synchronously is approximately proportional to their number, and, for those firing nonsynchronously, is approximately proportional to square root of their number (Blinowska and Durka 2006). The problem of the origins of the EEG rhythmical activity has been studied electrophysiological field when disserting about the brain nerve cells and when modeling of electrical activity of the neural populations (Niedermeyer et al. 2011). The question is as follows: Are the rhythms caused by single cells with pacemaker properties or by the oscillating neural networks? With regard of this, it has been shown that some thalamic neurons display oscillatory behavior, even in the absence of synaptic input (Jahnsen and Llinás 1984), since there is evidence that the intrinsic oscillatory properties of some neurons contribute to the shaping of the rhythmic behavior of networks to which they belong. However, these properties may not be sufficient to account for the network rhythmic behavior: it is generally accepted that cooperative properties of networks consisting of excitatory and inhibitory neurons connected by feedback loops play the crucial role in establishing EEG rhythms. The frequency of oscillation depends on the intrinsic membrane properties, on the individual neurons' membrane potential, and on the synaptic interactions' strength of. Bursts of oscillatory activity may constitute a mechanism by which the brain can regulate changes of state in selected neuronal networks and change the route of information (Niedermeyer et al. 2011). EEG is usually registered by means of electrodes placed on the brain scalp. They can be secured by an adhesive (like *collodion*) or embedded in a special snug cap. The resistance of the connection should be less than 10 (kΩ), so the recording site is first cleaned with diluted alcohol, and then a conductive electrode paste is applied to the electrode cup. Knowing the exact positions of the electrodes is very important for interpreting a single recording as well as comparing the results, hence the need for standardization. The traditional *10–20 electrode system* (Jasper 1958) states positions of 19 EEG electrodes (and two electrodes placed on earlobes A1/A2) related to the specific anatomic landmarks, such that 10–20% of the distance between them is used as the electrode interval. The first part of derivation name indexes the array's row—from the front of the head: Fp, F, C, P, and O. The second part is formed by numbers that are even on the left and odd on the right side, while in the center there is "z" or "0". Progress in topographic representation of EEG recordings brought demand for a larger amount of derivations. Electrode sites halfway between those defined by the standard 10–20 system were introduced in the extended 10–20 system (Pivik et al. 1993). EEG is a measure of potential difference; in the referential (or unipolar) setup, it is measured relative to the same electrode for all derivations. This reference electrode is usually placed on the earlobe, nose, mastoid, chin, neck, or scalp center. No universal consent exists regarding the best position of the reference electrode, because currents coming from bioelectric activity of muscles, heart, or brain propagate all over the human body. In the bipolar setup (montage), each channel registers the potential difference between two particular scalp electrodes. Data recorded in a referential setup can be transformed into any bipolar montage. The common "average reference" montage can be obtained by subtracting from each channel the average activity from all the remaining derivations. The *Hjorth* transform references each electrode to the four closest neighbors, which is an approximation of the *Laplace transform* (LT). LT is calculated as a second spatial derivative of a signal, offering information about vertical current density. For the best performance, it needs an adequate spatial sampling–interelectrode distance around 20 (mm) (e.g. 128 electrodes on the scalp).

The estimates obtained by means of LT for the electrodes lying at the scalp periphery are biased and have to be excluded. Contrary to the open question of the reference, the necessity of artifact rejections is universally acknowledged: *artifacts* are recorded signals that are non-cerebral in origin. They may be divided into one of two categories, depending on their origin: *physiological artifacts* or *non-physiological artifacts*. Physiological artifacts can stem from muscles or heart activity (EMG, ECG), eye movement (EOG), external electromagnetic field, poor electrode contact, and participant's movement. Corresponding signals (EMG, EOG, ECG, and body movements), registered simultaneously with EEG, could be helpful in the visual rejection of artifact-contaminated epochs. Non-physiological artifacts arise from two main sources: external electrical interference (power lines or electrical equipment), and internal electrical malfunctioning of the recording system (electrodes, cables, amplifier).

Furthermore, artifacts may reduce the performance of machine learning techniques. Several ways of handling physiological artifacts can be found in the literature. Artifacts may be avoided, rejected, or removed from the EEG dataset. Artifacts avoidance involves asking users to avoid blinking or moving their body during the experiments (Vigário 1997). This approach is very simple, because it does not require any computation, as brain signals are assumed to have no artifacts. However, this assumption is not feasible in an operational environment, since some artifacts, eye and body movements, are not easily avoidable. Artifact rejection approaches suggest to discard the epochs contaminated by the artifacts. Manual artifact rejection is an option to remove artifacts in brain signals, and experts could identify and eliminate all artifact-contaminated EEG epochs. The main disadvantage in using manual rejection is that it requires an intensive human labor, so this approach is not suitable for real-time evaluations. In the EEG, the following frequency rhythms are considered characteristic for its analysis (Fig. 2.6): delta [0.5–4 (Hz)], theta [4–8 (Hz)], alpha [8–12 (Hz)], beta [12–30 (Hz)], and gamma [above 30 (Hz)].

Delta activity is characterized by high amplitude and low frequency. It is usually associated with the slow wave in psychophysiology of sleep. It is suggested that it represents the onset of deep-sleep phases in healthy adults (Moser et al. 2009). Theta rhythm is generally linked to the hippocampus activity (Buzsáki 2005) as well as to the neocortex (Cantero et al. 2003). It is commonly believed to be linked to deep relaxation or meditation (Craigmyle 2013), and it has been observed during the transition between wake and sleep (Hagemann 2008). However, theta rhythms are suggested to be important for learning and memory functions (Sammer et al. 2006), encoding and retrieval (Ward 2003), which involve high concentration (Hagemann 2008). It has also been suggested that theta oscillations are associated with the attentional control mechanism in the anterior cingulate cortex (Craigmyle 2013; Smith et al. 2001), and it is often shown to increase with a higher cognitive task demand (Gundel and Wilson 1992). Alpha activity has been found in the visual cortex (occipital lobe) during periods of relaxation or idling (eyes closed but individual awake). In the continuous EEG, alpha band is characterized by high amplitude and regular oscillations, in particular over parietal and occipital areas.

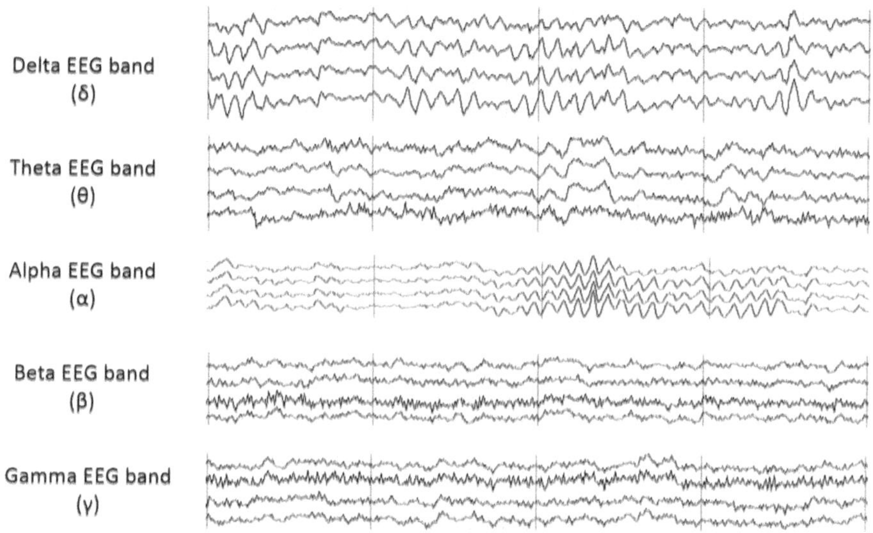


**Fig. 2.6.** Characteristic EEG rhythms, from the top: δ [0.5–4 (Hz)], θ [4–8 (Hz)], α [8–12 (Hz)], β [12–30 (Hz)]. The gamma band could reach 100 (Hz)

High alpha power has been assumed to reflect a state of relaxation or cortical idling; however, when the operator assigns more effort to the task, different regions of the cortex may be recruited in the transient function network leading to passive oscillation of the local alpha generators, in synchrony with a reduction in alpha power (Smith et al. 2001). Recent results have suggested that alpha is involved in auditory attention processes and in the inhibition of irrelevant tasks' areas, so as to enhance the signal-to-noise ratio (Gevins et al. 1998; Klimesch et al. 2007). Additionally, alpha activity may be further divided into subbands by means of the frequency corresponding to the user's alpha peak (Klimesch 1999), called *Individual Alpha Frequency* (IAF). For instance, *alpha 3* [$\mathrm{IAF} \div \mathrm{IAF} + 2$ (Hz)] reflects the semantic memory performance, while *alpha 1* and *alpha 2* [respectively, $\mathrm{IAF} - 4 \div \mathrm{IAF} - 2$ and $\mathrm{IAF} - 2 \div \mathrm{IAF}$ (Hz)] reflect general task demands and attentional processes. Beta activity is predominant in wakefulness state, especially in frontal and central areas of the brain. High power in beta band is associated with the increased mental arousal and activity. Dooley (2009) pointed out that a beta wave represents cognitive consciousness and active, business, or anxious thinking. This band can be further divided into *low beta wave* [12.5–15 (Hz)], *middle beta wave* [15–18 (Hz)], and *high beta wave* (>18 Hz). Low waves seem to be associated with inhibition of phasic movements during the sleep, and high waves with dopaminergic system (Hagemann 2008). Finally, gamma is the fastest activity in EEG and it is thought to be infrequent during waking states of consciousness (Dooley 2009). Recent studies reveal that it is linked with many cognitive functions, such as attention, learning, and memory (Jensen et al. 2007; Toppi et al. 2014). Gamma components are difficult to be recorded by the scalp electrodes, because of their low amplitude, but with *Electrocorticography* (ECoG) components up to 100 (Hz), or even higher, it might be registered.

The contribution of different rhythms to the EEG depends on the age, on the psycho-cognitive state of the participant, and on level of alertness. Considerable intersubject differences in EEG characteristics also exist, since EEG pattern is influenced by neuropathological conditions, metabolic disorders, and drug action (Niedermeyer et al. 2011).

## References

Blinowska K, Durka P (2006) Electroencephalography (EEG), Wiley encyclopedia of biomedical engineering. John Wiley & Sons, Inc., New Jersey. doi:10.1002/9780471740360.ebs0418

Buzsáki G (2005) Theta rhythm of navigation: link between path integration and landmark navigation, episodic and semantic memory. Hippocampus 15:827–840. doi:10.1002/hipo.20113

Cantero JL, Atienza M, Stickgold R, Kahana MJ, Madsen JR, Kocsis B (2003) Sleep-dependent theta oscillations in the human hippocampus and neocortex. J Neurosci Off J Soc Neurosci 23:10897–10903

Craigmyle NA (2013) The beneficial effects of meditation: contribution of the anterior cingulate and locus coeruleus. Front Psychol 4. doi:10.3389/fpsyg.2013.00731

Dooley C (2009) The impact of meditative practices on physiology and neurology: a review of the literature. Sci Discipulorum 4

Gevins A, Smith ME, Leong H, McEvoy L, Whitfield S, Du R, Rush G (1998) Monitoring working memory load during computer-based tasks with EEG pattern recognition methods. Hum Factors 40:79–91

Gundel A, Wilson GF (1992) Topographical changes in the ongoing EEG related to the difficulty of mental tasks. Brain Topogr 5:17–25. doi:10.1007/BF01129966

Hagemann K (2008) The alpha band as an electrophysiological indicator for internalized attention and high mental workload in real traffic driving [WWW Document]. http://docserv.uni-duesseldorf.de/servlets/DocumentServlet?id=8318. Accessed 1 Aug 2014

Jahnsen H, Llinás R (1984) Electrophysiological properties of guinea-pig thalamic neurones: an in vitro study. J Physiol 349:205–226

Jasper H (1958) The ten twenty electrode system of the international federation. Electroencephalogr Clin Neurophysiol 10:371–375

Jensen O, Kaiser J, Lachaux J-P (2007) Human gamma-frequency oscillations associated with attention and memory. Trends Neurosci 30:317–324. doi:10.1016/j.tins.2007.05.001

Klimesch W (1999) EEG alpha and theta oscillations reflect cognitive and memory performance: a review and analysis. Brain Res Rev 29:169–195. doi:10.1016/S0165-0173(98)00056-3

Klimesch W, Sauseng P, Hanslmayr S (2007) EEG alpha oscillations: the inhibition-timing hypothesis. Brain Res Rev 53:63–88. doi:10.1016/j.brainresrev.2006.06.003

Moser D, Anderer P, Gruber G, Parapatics S, Loretz E, Boeck M, Kloesch G, Heller E, Schmidt A, Danker-Hopfe H, Saletu B, Zeitlhofer J, Dorffner G (2009) Sleep classification according to AASM and Rechtschaffen & Kales: effects on sleep scoring parameters. Sleep 32:139–149

Niedermeyer E, Schomer DL, Lopes da Silva FH (2011) Niedermeyer's electroencephalography: basic principles, clinical applications, and related fields. Wolters Kluwer/Lippincott Williams & Wilkins Health, Philadelphia

Noback CR, Strominger NL, Demarest RJ, Ruggiero DA (2005) The human nervous system: structure and function. Springer Science & Business Media

Pivik RT, Broughton RJ, Coppola R, Davidson RJ, Fox N, Nuwer MR (1993) Guidelines for the recording and quantitative analysis of electroencephalographic activity in research contexts. Psychophysiology 30:547–558

Sammer G, Reuter I, Hullmann K, Kaps M, Vaitl D (2006) Training of executive functions in Parkinson's disease. J Neurol Sci 248:115–119. doi:10.1016/j.jns.2006.05.028

Smith ME, Gevins A, Brown H, Karnik A, Du R (2001) Monitoring task loading with multivariate EEG measures during complex forms of human-computer interaction. Hum Factors 43: 366–380

Toppi J, Mattia D, Anzolin A, Risetti M, Petti M, Cincotti F, Babiloni F, Astolfi L (2014) Time varying effective connectivity for describing brain network changes induced by a memory rehabilitation treatment. In: Presented at the 2014 36th annual international conference of the IEEE Engineering in Medicine and Biology Society, pp 6786–6789. doi:10.1109/EMBC.2014.6945186

Vigário RN (1997) Extraction of ocular artefacts from EEG using independent component analysis. Electroencephalogr Clin Neurophysiol 103:395–404

Ward LM (2003) Synchronous neural oscillations and cognitive processes. Trends Cogn Sci 7:553–559

# Chapter 3
# Mental States in Aviation

This chapter starts with the first neurophysiological study performed in aviation, dated back to 1878, and it represents the birth of the Aviation Medicine. The following three paragraphs provide the state of the art of neurophysiological and psychophysiological works concerning the most investigated and important mental states in aviation. Specifically, the mental workload, cognitive control behavior, and training will be initially described from a *human factor* point of view. Then, they will be characterized from cognitive processes perspectives, and finally in terms of neurophysiological variations. As the reader will see, those mental states are not completely independent each other. For example, before the evaluation of the mental workload it is very important to assess the training level of the user in order to avoid results obtained due to learning progresses, and not due to different workload demands. The objective analysis of such mental states allows to improve human–machine interactions, to better design human–computer interfaces, to tailor training programs, and to adaptively support users during the execution of their works with the aim to avoid unpaired mental states and, consequently, enhance error commission probability.

## 3.1 Aviation Medicine Hystory

The nineteenth century led to continuous progress in scientific understanding within aviation environments. No individual had a greater impact on the incorporation of experimentation and scientific principles in medicine than French physiologist, Dr. Claude Bernard (1813–1878). His scientific feats were extraordinary in scope and large in number, and many of his rigorous rules for the application of scientific ideals remain as true today as when he wrote them over a hundred years ago. Using

 
G. Borghini et al., *Industrial Neuroscience in Aviation*,
Biosystems & Biorobotics 18, DOI 10.1007/978-3-319-58598-7_3

objective scientific experimentations instead of conventional wisdom and tradition, Dr. Bernard discovered the physiological functions of multiple organ systems and first coined the term *homeostasis*. In 1865, he published a discourse on the scientific method, which he titled as *An Introduction to the Study of Experimental Medicine* (Bernard 1957). Many of the themes familiar to any modern day scientific researcher were first introduced by Bernard in this publication, for example, the effects of carbon dioxide, oxygen, and hemoglobin on pilots.

Another French physiologist, Paul Bert, proted of Claude Bernard, later introduced vast amounts of information into the new field of flight physiology, which has earned him the epithet of "Father of Aviation Medicine". Dr. Bert was very interested in the effects of altitude on human physiology and conducted several experiments to research and validate his hypotheses. Not content with experimenting solely with balloonists, he developed the first hypobaric chamber, which was able to simulate altitudes up to 36,000 ft. He experimented with animals to determine the minimum required partial pressure of oxygen in circulating blood. Using the results from these experiments, he came to the conclusion that the use of supplemental oxygen in high altitude balloon travel was necessary, and fervently urged operators to do so. He discovered oxygen toxicity. Ultimately, Dr. Bert compiled all of his research into a simple book titled *La Pression Barometrique* (Bert 1878), which was published in 1878. This book would form much of the early foundation for the budding discipline of Aviation Medicine.

## 3.2 Mental Workload

The workload is a complex construct that is assumed to be reflective of an individual's level of attentional engagement and mental effort (Wickens 1984). Measurement of mental workload essentially represents the quantification of mental activity resulting from performing a task—or a set of tasks. Several empirical investigations have indicated that performance declines at the extremes of the workload demand continuum—that is when the event rate is excessively high or extremely low (Borghini et al. 2014b). Consequently, it is important to preserve a good level of the user's mental workload, avoiding either mental underload or overload state, with the aim to keep an optimal level of performance and reducing the probability of errors commission (Parasuraman and Hancock 2001). For these reasons, the mental workload is an important and central construct in ergonomics and human factor researches.

Various mental workload definitions have been given during the last decades:

- Mental workload refers to the portion of operator information processing capacity or resources that is actually required to meet system demands (Eggemeier et al. 1991);
- Workload is not an inherent property, but rather it emerges from the interaction between the requirements of a task, the circumstances under which it is performed, and the skills, behaviors, and perceptions of the operator (Hart and Wickens 1990);

- Mental workload is a hypothetical construct that describes the extent to which the cognitive resources required to perform a task have been actively engaged by the operator (Gopher and Donchin 1986);
- The reason to specify and evaluating the mental workload is to quantify the mental cost involved during task performance in order to predict operator and system performance (Cain 2007).

Such definitions show that the mental workload may not be a unitary concept because it is the result of different interacting aspects. In fact, several mental processes such as alertness, vigilance, mental effort, attention, stress, mental fatigue, drowsiness, and so on can be involved meanwhile in a task execution and they could be affected by specific tasks demand in each moment. In general, mental workload theory assumes that: (i) people have a limited cognitive and attentional capacity, (ii) different tasks will require different amounts (and perhaps different types) of processing resources, and (iii) two individuals might be able to perform a given task equally well, but differently in terms of brain activations (Baldwin 2003; Wickens et al. 2012).

In the last 20 years, it has been widely documented that the 70% of civil aviation accidents were linked with human factors (Bellenkes 2007). Recently, the Aviation Safety Network reported 37 accidents with 564 casualties, and 8 accidents with 295 casualties, respectively, during 2013 and the first months of 2014. Moreover, the air traffic is growing exponentially and it has been predicted to double in 2020 (Flight Safety Foundation). This factor would increase the work difficulty faced by pilots and air traffic controllers (ATCOs) and in selecting the requested aptitudes (Broach 2013; Broach et al. 1997). In particular, the latter ones have to perform a variety of tasks, including monitoring air traffic, anticipating loss of separation between aircrafts, and intervening to resolve conflicts and minimize disruption to air traffic flow (for an extensive compilation of the tasks and goals of *en route* control, see (Rodgers and Drechsler 1993). In this domain, the pilots and ATCOs' behavior could already be measured through several human factor tools, such as the explicit measurement of errors committed during the execution of the task, or by using questionnaires related to the subjective workload perception, such as the *Instantaneous Self-Assessment* (ISA—Tattersall and Foord 1996), *NASA—Task Load Index* (NASA-TLX—Hart and Staveland 1988) or the *Subjective Workload Assessment Technique* (SWAT—Reid and Nygren 1988) questionnaires. Because of their inherently subjective nature, none of such questionnaires allows to have an objective and reliable measure of the actual cognitive demand in a real environment. All the described questionnaires have pros and cons, but there is not a generally accepted standard (Rubio et al. 2004). Therefore, the need of an objective measure became fundamental for reliable workload evaluations.

Many neurophysiological measures have been used for the mental workload evaluation, including *electroencephalography* (EEG), *functional near-infrared* (fNIR) imaging (Ayaz et al. 2012, 2013; Dehais et al. 2010; Durantin et al. 2014; Roy et al. 2016), *functional Magnetic Resonance Imaging* (fMRI), *magnetoencephalography* (MEG), and other types of biosignals such as *electrocardiography* (ECG), *Electrooculography*, (EOG) and *galvanic skin response* (GSR) (Borghini et al. 2012; Duru et al. 2013; Ryu and Myung 2005; Warm et al. 2008). The size, weight, and power constraints outlined in the introduction limit the use of some techniques in operational environments. In fact, fMRI and MEG techniques require room-size equipment that are not portable as appropriate structures like electro-magnetic interference shields (Cabeza and Nyberg 2000). EOG, ECG, and GSR activity measurements highlighted correlations with some mental states, but they have been demonstrated to be useful only in combination with measures directly linked to the brain activity (Borghini et al. 2014a, b; Ryu and Myung 2005). Consequently, the EEG and fNIR are the most likely candidates that can be straightforwardly employed in operational environments. According to the idea that the higher the mental workload level is, the greater the brain blood oxygenation will be, the *functional near-infrared spectroscopy* (fNIRs) has also been demonstrated to be another reliable mental workload measurement technique (Cui et al. 2011; Derosière et al. 2013). fNIR spectroscopy is safe, highly portable, user-friendly, and relatively inexpensive, with rapid application times and near-zero run-time costs, so it could be a potential portable system for measuring mental workload under realistic settings. The most common fNIR system uses infrared light introduced in the scalp to measure changes in blood oxygenation. *Oxyhemoglobin* ($HbO_2$) converts to deoxyhemoglobin (HbR) during neural activity that is the cerebral hemodynamic response. This phenomenon is called *blood-oxygen-level dependent* (BOLD) signal. fNIRs has been shown to compare favorably with other functional imaging methods (Huppert et al. 2006) and demonstrates solid *test–retest* reliability for task-specific brain activation (Herff et al. 2013; Plichta et al. 2006). Thus, the primary hypothesis was that blood oxygenation in the prefrontal cortex, as assessed by fNIR, would rise with increasing task load and would demonstrate a positive correlation with the mental workload. In fact, Izzetoglu et al. (2004, 2015) clearly indicated that the rate of changes in blood oxygenation was significantly sensitive to task load variations. Regarding the EEG measurements, most part of the studies show that the brain electrical activities mainly considered for the mental workload evaluation are the theta and alpha EEG rhythms, in particular on the *prefrontal cortex* (PFC) and the *posterior parietal cortex* (PPC) regions. The theta rhythm, especially over the PFC, presents a positive correlation—i.e., increases when the mental workload increases (Borghini et al. 2013; Gevins and Smith 2003; Smith and Gevins 2005), while the alpha rhythm, especially over the PPC, presents an inverse correlation—i.e., decreases (Brookings et al. 1996; Gevins et al. 1997; Jaušovec and Jaušovec 2012; Klimesch et al. 1997; Venables and Fairclough 2009). Only few studies have reported results about other EEG bands—i.e., the delta, beta, and gamma EEG rhythms. Onton et al. (2005) reported that the frontal midline theta rhythm increases with memory load, confirming previous results about the correlation between the frontal theta EEG activity and the mental effort (Gevins et al. 1997; Smith and Gevins 2005). Mental workload is also known for suppressing the EEG alpha rhythm and increasing the theta rhythm during the activity of encoding and retrieving information (Klimesch 1999; Klimesch et al. 1997). Several researches in the ATM domain treated the neurophysiological measurements of ATCOs' mental workload in realistic settings with the aim of developing HMI systems. These studies were based on both EEG and fNIRs techniques. It has been discussed how each technique is able to provide reliable estimation of mental workload. The propensity in using EEG or fNIRs techniques in such kind of applications has not been clarified yet. In fact, there are several factors about real operational scenarios to be taken into account. For example, both EEG and *fast optical signal* (FOS)-based fNIR have similar bandwidth and sample rate requirements, as the FOS appears to directly reflect aggregated neural spike activity in real time and can be used as a high-bandwidth signal akin to EEG (Medvedev et al. 2008). However, EEG and fNIRs systems have different physical interfaces, sizes, weights, and power budgets, thus different wearability and usability in real operational contexts. Specifically, the physical interface merits scrutiny, as it is non-trivial to maintain a good contact between the sensors (i.e., electrodes or optodes) and the brain scalp in freely moving tasks. The use of fNIRs eliminates motion artifacts and the need of both scalp abrasion and conductive gel. In addition, there is not the necessity to wear a cap but only a headband. Furthermore, unlike EEG, fNIRs recordings are not affected by electroculographic or facial electromyographic activity and environmental electrical noise, which are undoubtedly ubiquitous in human–computer interactions. Thus, this technology could appear more suitable in realistic environments (Goldberg et al. 2011). In two recent studies (Ayaz et al. 2012, 2013), the use of fNIRs technique to assess the mental workload has been investigated in the ATM environment. Based on the hypothesis (Durantin et al. 2014; Izzetoglu et al. 2004; Owen et al. 2005) that the hemodynamic response over the dorsolateral and ventrolateral prefrontal cortex was positively responsive to mental workload, Ayaz et al. (2012) demonstrated how the BOLD signal was a reliable workload indicator. In fact, by using this index, it was possible to discriminate significantly the different workload requested by two different kinds of communication (Data and Voice) on 24 professional ATCOs. Furthermore, (Ayaz et al. 2013; Bunce et al. 2011) showed that fNIR could be successfully used in realistic environments to assess: (i) mental workload levels of ATCOs performing standardized *N-back* task and operative ATM tasks, and (ii) expertise development when learning a complex cognitive and visuomotor tasks on *unmanned aerial vehicles* (UAV) pilots. However, in a recent study, Harrison et al. (2014) reported how the BOLD signal showed a lower resolution than the subjective measures (e.g., ISA) to evaluate the mental workload of ATCOs involved in the experiment. In particular, while the task was becoming more difficult, the subjective measure was still increasing, and the BOLD signal (neurophysiological index) reached its maximum, lingering on this value. Furthermore, the BOLD signal, used as workload index, was shown to be not reliable over time since the workload measurements performed in different days were significantly different and in discordance with the subjective measures. Indeed, also this technique has its weaknesses. Since the presence of hair may impact on both the photon absorption (Murkin and Arango 2009) and the coupling of the probes with the underlying scalp, the fNIRs technique is very reliable only on those unhairy brain areas, like the PFC. As quoted above, also the parietal brain sites play a key role in the mental workload evaluation, and (Derosière et al. 2013) pointed out how some fNIRs-measured hemodynamic variables were relatively insensitive to certain changes in mental workload and attentional states. Due to its higher temporal resolution and usability, in comparison with the fNIRs technique, the EEG technique overcomes such kind of issues.

In addition, there are several studies in ATM domain that highlighted the high reliability of EEG-based mental workload indexes (Brookings et al. 1996). The results showed that the effects of the task demand were evident on the EEG rhythms variations. EEG power spectra increased in the theta band, while significantly decreased in the alpha band as the task difficulty increased, over central, parietal, frontal, and temporal brain sites. In a recent study, (Shou et al. 2012) evaluated the mental workload during an ATC experiment using a new *time-frequency independent component analysis* (tfICA) method for the analysis of the EEG signal. They found that "the frontal theta EEG activity was a sensitive and reliable metric to assess workload and time-on-task effect during an ATC task at the resolution of minute(s)." In other recent studies involving professional and trainees ATCOs (Aricò et al. 2015; Borghini et al. 2014a; Di Flumeri et al. 2015), it has been demonstrated how it was possible to compute, by machine learning techniques and specific brain features, an EEG-based workload index which is able to significantly discriminate the workload demands of the ATM task. In those studies, the ATM task was developed with a continuously varying difficulty levels in order to ensure realistic ATC conditions, i.e., starting from an easy level, then increasing up to a hard one and finishing with an easy one again. The EEG-based mental workload index shows to be directly and significantly correlated with the actual mental demand experienced by the ATCOs during the entire task. The same EEG-based workload index was also used to evaluate and compare the impact of different avionic technologies on the mental workload of professional helicopter pilots (Borghini et al. 2015). Furthermore, the machine learning techniques have been successfully used in other real environments for the evaluation of mental states (Müller et al. 2008) and mental workload (Berka et al. 2004). Even if the main limitation of the EEG is its wearability, technology improvements (Liao et al. 2012) have being developed and tested in terms of dry electrodes (no gel and problems of lowering the impedances), comfort, ergonomic, and wireless communications (no cables between the EEG cap and the recording system).

In conclusion, the EEG technology seems to be the most appropriate solution to evaluate the mental workload in realistic and operational settings, and to be integrated in passive BCI systems. Such systems will support the operator during his/her working activity in order to improve the works wellness and, most of all, the safety standards of the whole environment.

*Workload and Autonomic Signals: Heart and Eyes Activities*

*Heart rate* (HR) has been demonstrated to be a reliable measure of mental workload. Generally, there is a direct correlation between workload and heart rate, so when the first increases the second increases as well (Costa 1993; Hankins and Wilson 1998; Jorna 1992; Roscoe 1993; Veltman and Gaillard 1996; Wilson et al. 1994, 1999). Nevertheless, some authors are critical toward the use of HR for evaluating workload because of the various psychological, environmental, and emotional factors that can influence the response (Jorna 1992; Lee and Park 1990; Roscoe 1993). For example, uncertainty and anxiety can significantly raise the heart rate (Jorna 1993). Gravitational forces may also affect this technique (Wilson 1993). Furthermore, heart rate does not measure absolute levels of workload, but relative levels: This may be a benefit in real world rather than simulated conditions because there is less control over the variables (Roscoe 1992, 1993). A study about multitasking performance (Fairclough et al. 2005) explored the interaction between learning and task demand on psychophysiological reactivity. These authors used EEG activity, cardiac activity, and respiration rate to evaluate the impact of task demand and learning, and found out that the sustained response to task demand was characterized by a reduction of parasympathetic inhibition (reduced vagal tone and increased heart rate), as well as a reduced eye blink duration. This sustained response may represent the state of focused concentration that was necessary to perform a multitasking activity.

In 1993, Wilson evaluated ten pilots who were required to fly 90-min flight scenario to test the reliability of psychophysiological measures of the workload. Cardiac, electrodermal, and electrical brain activity's measures were highly correlated and exhibited changes in response to the demands of the flight phases (Wilson 1993).

Using the eye activity measurement, Wilson found that blink rates decreased significantly during the more highly visually demanding segments of the flights.

Another important technique to assess the mental workload is the analysis of the *eye blinks rate* (EBR). In effect, eye blinks and blink duration decrease when visual workload intensifies (Brookings et al. 1996; De Waard 1996; East 2000; Hankins and Wilson 1998; Van Orden et al. 1999). It has been shown that, although ocular measures are sensitive to mental demands, they are affected by other factors, such as fatigue, body or head moves, and environmental changes. A change in light or air quality may influence eye blinks rate (De Waard 1996). Some studies (Sirevaag et al. 1993) presented contradictory results caused by different methods used for the experiments. Furthermore, as reported by Cain (2007), blinks measures can be context-dependent. Blink rate decreases with increased workload with the processing of visual stimuli, while increases with the augment of workload in memory tasks (Wilson 1993). Effectiveness, as in the connection between blinks rate and workload, seems tenuous, according to (Castor et al. 2003). Blink closure duration decreases with increasing of visual stimuli workload or acquisition of data from a wide field of view, while blink latency increases with memory and response demands (Castor et al. 2003). Stern and Skelly (1984) found that blinking is "inhibited during the acquisition of information, whether such information is presented visually or auditory. Once a decision is made whether it requires action or requires the inhibition of action, a blink is likely to occur. The non-inhibition of blinking during that time periods is associated with a higher probability of occurrence of missed signals and erroneous responses." In the field of flight and driving research, some authors tried to discriminate between visual and mental workload to make the eye blinks rate (Van Orden et al. 1999) and eye blinks duration (Sirevaag et al. 1993) a good index of estimation in some aspects of mental workload

(Miller 2001). Some authors found that eye blinks is a good measure for visual workload (Brookings et al. 1996; East 2000; Hankins and Wilson 1998) as the eye blinks rate decreased significantly when the task became more difficult.

## 3.3 Cognitive Control Behavior

Training can result not only in the acquisition of new skills (Hill and Lent 2006; Satterfield and Hughes 2007) but also in improved declarative knowledge and enhanced strategic knowledge—defined as knowing when to apply a specific procedure or skill especially during unexpected events (Kozlowski et al. 2001). Furthermore, despite the time passed from the last training session, there might be the need to assess if the operator is still able to work ensuring high performance levels, hence, a proper level of safety. For such a reason, another issue is the necessity of objectively monitoring and assessing the operators' performance (Leape and Fromson 2006) in terms of cognitive control behaviors. Nevertheless, although the results in terms of performance should be the same, the cognitive demand could differ over the time for the same operator. In other words, after a certain period, the operator should be still able to execute the considered task by achieving the same performance level, but it might require a different amount of cognitive resources. Therefore, different operators could achieve the same performance results but involving different amount of cognitive resources, thus showing different expertises. Nowadays, there are no tools providing objective and quantitative information concerning expertise levels and aimed to better manage the training program and personnel selection. As a reference, the SRK framework introduced by Rasmussen (1983) was considered and still used in the aeronautic field. The aim of this framework is to explain human behavior and describe the wide range of mental capabilities used in everyday situations (e.g., working environments).

During recent decades, many attempts have been made to extend these models to higher level human decision-making to conform with the increasing levels of automation in operational environments (e.g., aviation, hospitals, public transport), and to transfer such models for process control applications. In this effort, it has to be considered that humans are not simply deterministic *input–output* devices, but *goal-oriented* creatures who actively select their objectives and seek the relevant information in order to fulfill them (Polanyi 2012; Rosenblueth et al. 1943). At a higher level of conscious planning, most human activity depends upon a rather complex sequence of activities, and feedback correction during the course of behavior from mismatch between goal and final outcome will therefore be too inefficient, since in many cases it would lead to a strategy of blind search (Rasmussen 1986; Rasmussen and Vicente 1989). Human activity in a familiar environment will not be goal-controlled; rather, it will be oriented toward the goal and controlled by a set of rules that, previously, has proven to be successful. In unfamiliar situations when proven rules are not available, the behavior may be goal-controlled in the sense that different attempts are made to reach the goal, and a successful sequence is then selected. The efficiency of humans in coping with complexity is largely due to the availability of a large *repertoire* of different mental representations of the environment, from which rules to control behavior can be generated *ad hoc.* When we distinguish categories of human behavior according to basically different ways of representing the constraints in the behavior of a deterministic environment or system, three typical levels of performance emerge: *skill-*, *rule-*, and *knowledge*-based performance (Rasmussen 1983). These levels and a simplified illustration of their interrelation are shown in Fig. 3.1. The *skill-based behavior* represents the sensory motor performance during acts or activities that, following an intention statement, take place without conscious control, as smooth, automated, and highly integrated patterns of behavior. In most skilled sensory-motor tasks, the body acts as a multivariable continuous control system, which synchronizes movements according to the behavior of the environment. Performance is based on feedforward control and depends upon a very flexible and efficient dynamic internal world model. Characteristically, skilled performance rolls along without conscious attention or control. The total performance is smooth and integrated, and sense input is not selected or observed: the senses are only directed toward the aspects of the environment needed subconsciously to update and orient the internal map. In general, human activities can be considered as a sequence of such skilled acts or activities composed for the actual occasion. The flexibility of skilled performance is due to the ability to compose, from a large *repertoire* of automated subroutines, the sets suited for specific purposes (Fig. 3.1).

At the next level of *rule-based behavior,* the composition of such a sequence of subroutines in a familiar work situation is typically controlled by stored rules or procedures, which may have been derived empirically during previous occasions or may be prepared for that occasion through conscious problem-solving and planning processes. In this case, the performance is goal-oriented but structured through a stored rule by the "feed forward control". Very often, the goal is not even explicitly formulated, but it is implicitly found in the situation releasing the stored rules, that is, selected from previous successful experiences. Feedback correction during the performance will require functional understanding and analysis of the current response of the environment, which may be considered as an independent concurrent activity at the next higher level (*knowledge-based*). The boundary between skill-based and rule-based performance is not quite distinct, and much depends on the level of training and on the attention of the person. In general, the skill-based performance rolls along, without the person's conscious attention, and he/she will be unable to describe how he/she controls and on what information he/she bases the performance. The higher level, the *rule-based coordination*, is generally based on an explicit know-how, and the person can report the rules used: during unfamiliar situations, faced with an environment for which no know-how or rules for control are available from previous encounters, the control of performance must move to a higher conceptual level, in which performance is goal-controlled and *knowledge-based.* In this situation, the goal is explicitly formulated, based on an analysis of the environment and the person's overall aims: after having considered different plans

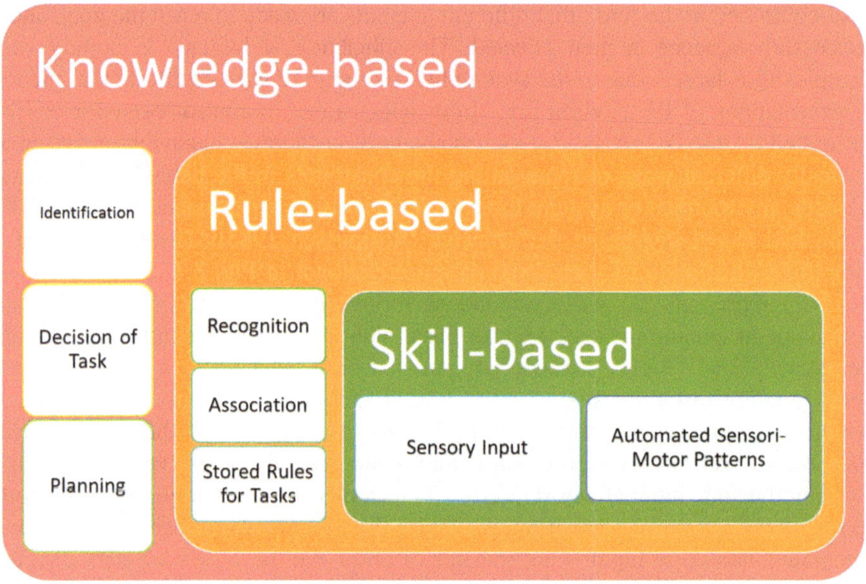


**Fig. 3.1** Simplified illustration of three levels of the human performance model proposed by Rasmussen. Note that skill, rule, and knowledge levels are not alternatives, but interact in a complex way—only rudimentarily represented in the diagram

and having selected them by testing their efficacy with regard to the considered goal, the person develops a suitable one. The above-mentioned selective process can occur physically—by trial and error—or conceptually—by means of understanding the functional properties of the environment and prediction of the effects of the considered plan. At this level of functional reasoning, there is a "mental model"—which may take several different forms—that explicitly represents the internal structure of the system. An example is learning to use the T9-keyboard system on the mobile phone: the first time that it happens, this learning process is knowledge-based, since it diagnoses the way for using a T9-keyboard system and produce a rule. When the procedure to select letters and words is already known, the control shifts to the rule-based behavior for applying the rules learned. Finally, after practice, the procedure may turn automated, therefore skill-based. Similar distinctions between different categories of cognitive control behavior have been proposed elsewhere. For example, Fitts and Posner (1979) distinguished between three phases of learning a skill: the early or cognitive phase, the intermediate or associative phase, and the final or autonomous phase. If we consider that in real life, a person will have a varying degree of training when performing a task depending on variations and disturbances, the correspondence with the three levels in the present context is clear. One aspect of the categorization of human performance in skill/rule/knowledge-based behavior is the role of the information observed from the environment, which is basically different in the different categories. This is the case even though major problems during unfamiliar situations may be caused by the fact that the same indication may be perceived in various different roles and that it is a well-known psychological phenomenon that shifts between different modes of perception is difficult (Rasmussen 1986).

*Toward a Possible Integration Between SRK and Neurophysiological Variables*

Several interesting issues about the neuroanatomical and functional models may be related to the level of attentional control of the SRK framework, for a possible integration between SRK and neurophysiological variables. One of the first models of attentional control was proposed by Posner and Petersen (1990) and consists of two systems that are separated but also closely interrelated: an anterior attentional system (linked to the environmental monitoring and detection of target stimuli) and a posterior attentional system (linked to the orientation of the attention). More recent versions of this model describe three systems (Kochanska et al. 2000; Posner and Rothbart 2007; Raz and Buhle 2006):

- *Alert/vigilance system* connected to the frontal and parietal regions (in particular of the right hemisphere) and aimed at maintaining a state of activation;
- *Orienting system* consists of the posterior parietal and the frontal cortex, the temporoparietal junction (TPJ), the thalamic nuclei—such as pulvinar and reticular nuclei—and the superior colliculus. Its functions include anchoring, dis-anchoring and shift of attention, as well as selection of specific information from multiple sensory stimuli;
- *Executive system* consists of the prefrontal medial cortex, the anterior cingulate cortex, and the supplementary motor area. Its functions include the voluntary control of behavior, the conscious elaboration of experience, the handling of new situations, the monitoring, and the resolution of conflicts. These conflicts may include planning or decision-making, error detection execution of new responses, inhibitory control, self-regulation, and involvement in stressful conditions. It is proposed that the *anterior cingulate cortex* (ACC) has a crucial role in monitoring the environment, detecting the target stimuli appearance and resolving conflicts between stimuli.

Furthermore, Corbetta and Shulman (2002), by incorporating several recent studies, have proposed that the cortical control of attention is divided into two functionally distinct but interacting systems:

- *Dorsal frontoparietal network* (top-down, endogenous attention) is guided by the cognition and is involved in control of goal-driven attention, preparation, and application of relevant stimuli selection. In the dorsal frontal regions, this system includes frontal eye fields (FEF) and medial and lateral frontal cortex while, in the dorsal parietal regions, it includes the intraparietal sulcus (IPS) and the superior parietal lobule (SPL).
- *Ventral frontoparietal network* (bottom-up, exogenous attention) is guided by perception and is a stimulus-driven attentional system that is involved in dis-engagement and reorienting of attention toward salient or unexpected stimuli. The attentional shifts are automated. The areas involved are the temporoparietal

junction (TPJ), the inferior parietal lobule (IPL), the superior temporal gyrus (STG), and the ventral frontal cortex (VFC), in particular the inferior and the middle frontal gyrus (IFG, MFG).

The SRK framework can be connected to the models mentioned above, which provide an anatomical and functional base to do a step forward the investigation of neurophysiological correlates of Rasmussen's taxonomy. In fact, among the cognitive neuropsychologists, in terms of limited capacity it has been hypothesized a mechanism aimed at programming and controlling the cognitive processes in relation to priorities, goals, and external conditions. This mechanism called *central executive* provides a higher level processing in a hierarchical organization. Since the central executive has a limited capacity, many routine operations should be delegated to mechanisms that operate automatically, regardless of the voluntary attention costs (in terms of effort). An automatic process (Shiffrin and Schneider 1977) consists in the activation, on the base of appropriate inputs, of a learned sequence of elements that proceeds without intentional attention (as in without the active control by the subject). Instead, a controlled process has a limited capacity, since it requires continuous effort and monitoring by the subject. It works in a serial mode, drawing on the stock of short-term memory. The automation (or proceduralization) is facilitated by practice, which makes the process faster, parallel, and less demanding for the mental load. For example: driving a car, where the processes that were previously "controlled", subsequently becomes largely automated because various operations are performed without the use of conscious attention. Anyway, in special cases (the use of a new car or writing a text in a foreign language), the return to control of different operations, that in usual conditions are automated, is needed. However, it should be remembered that attention does not always coincide with controlled process: the automatic process also focuses attention on certain stimuli, selecting them among others, deemed not relevant for the task, and keeps focusing on them for all the necessary time. Neumann (1984) pointed out that the automatic process is not completely uncontrolled, but that actually control is below the threshold of consciousness. Norman and Shallice (1986) distinguished between totally automatic and partially automated processes: the latter is based on selection (in the absence of conscious or voluntary control) between the established schemes, and the selection is based on immediate priorities dictated by learning or by the context. Thus, it seems appropriate to distinguish between *non-intentional* attentional processes—which involve minimal effort and self-perception of the subject, and *intentional attention*, conscious and limited capacity. The relationship between awareness and attention is therefore extremely complex. Several empirical studies (Kellogg 1980; Marcel 1983; Polich 1986) demonstrate how information receiving not consciously attention is processed in relation to its meaning, and how the latter can also be analyzed in absence of intentional control (and therefore of consciousness). Starting from the idea that the attentional mechanisms control the access to awareness (Baars 1997; LaBerge 1995), different models tend to identify the function of the attentional control with conscience (as a central operative system of activation and self-monitoring).

According to Allport (1988), Schmidt et al. (1990), and Sadaghiani et al. (2010) intentionality, attention, control, and awareness are the constitutive dimensions of consciousness. Other authors distinguish between consciousness of external reality and self-awareness or awareness of our own cognitive processes (metacognition), our emotions and our choices.

## 3.4 Training

Training could not only result in the acquisition of new skills (Hill and Lent 2006; Satterfield and Hughes 2007), but also in improved declarative and enhanced strategic knowledge—defined as knowing when to apply a specific procedure or skill, in particular during unexpected events (Kozlowski et al. 2001). Inappropriate training might have high social costs, especially for personnel that operate in complex operational environments, such as air traffic controllers or pilots. One of the current limitations of the standard training assessment procedures is the lack of objective information about the cognitive resources requested by the user for the correct execution of the proposed task across the different training sessions. The training level as well as the related performance are generally evaluated by expert supervisors, thing that makes this approach highly operator-dependent. In fact, even if a task is perfectly accomplished, by means of the standard assessment methods it will not be possible to estimate information about the user's cognitive capacity and expertise. Therefore, it becomes difficult to quantify the cognitive resources available to the user to accomplish unexpected events or emergency conditions. Thus, inappropriate training evaluation could bring to high probability of committing mistakes, and the consequences might have important social and economic impacts. In 1928, Spearman stated that psychological writings "crammed" a lot of allusions to *Human Factor* (HF) in an "incidental manner", but "they hardly arrived at considering such a concept in a systematic and profound way" (Spearman 1928). In the last decades, there has been an increasing interest in studying HFs in working environments, especially its causes and possible prevention strategies. For example, in the aviation field, investigations of aircraft accidents had revealed that 80% of the accidents are based on human errors, and further investigations indicated that a significant portion of human errors is attributable to HF failures primarily associated with inadequate communication and coordination of the crew (Taggart 1994). Several researches, moreover, have tried to seek solutions aimed at improving safety, and many of them have identified the personnel training as a significant strategy in achieving such a goal (Barach and Small 2000; Barach and Weingart 2004; Hamman 2004; Leonard et al. 2004; Leonard and Tarrant 2001). *Training* refers to a systematic approach to learning and development with the aim of improving individuals', teams', and organizations' effectiveness (Goldstein and Ford 2002). The importance and interest in the concept of training in operational environments (e.g., aviation, hospital and public transport) is reflected by the regular publications of scientific reviews in the *Annual Review of Psychology* since

1971 (Aguinis and Kraiger 2009; Campbell 1971; Goldstein 1980; Salas et al. 2001; Tannenbaum and Yukl 1992; Wexley 1984). An objective quantification of cognitive resources required by the operator during the execution of a task is very important, since the human brain resources are limited. The difference between the cognitive resources actually available and the amount of those involved for the task execution is called *cognitive spare capacity* (Rudner and Lunner 2014). The higher the cognitive spare capacity during a normal working activity (i.e., the operator is involving a low amount of cognitive resources), the greater is the operator's ability to perform secondary tasks or to react to unexpected—emergency events. It is a common experience that, when subjects face a task for the first time, they attempt to understand its requests in a trial and error fashion; eventually, the subjects gain experience up to the full comprehension of the task that it is quickly followed by an increase of their task-related performance. For example, when we drive a car for the first time we pay high attention to many details, but after a while we are able to drive the car much better than at the beginning, and we are also able to perform other tasks in parallel. It was shown that the learning is due to the acquisition and memorization of solutions to specific situations (Logan 1988), while individual information of general meaning can be grouped for creating more specific and larger knowledge by a practice (Newell and Rosenbloom 1993). In addition, it was sustained that the automation progress is due to two consecutive phenomena: at the beginning, the memorization of the procedures allows to substitute general knowledge with specific (procedural) knowledge; later, a decrease of the involvement of the cerebral attention system is observed (Anderson 1993) since the acquired procedures will become automatically performed. Such learning process is well described by the *learning curve* (Ritter and Schooler 2001), where the performance enhances and the time necessary for the task execution decreases along with the training sessions. A focused and progressive training program makes the people to gain experience and knowledge for facing and reaching appropriate standards of performance (Erickson et al. 2007; Gopher 1993; Kramer et al. 1999; Parasuraman and McKinley 2014). In addition, training programs in which performance on a task is methodically improved through adaptive performance feedback over multiple sessions (instructional strategies and individualized program) have been shown to reliably reduce costs associated with performing multiple concurrent tasks (Bherer et al. 2005; Glass et al. 2000; Kramer et al. 1995, 1999; Schumacher et al. 1999).

In this context, the possibility to objectively assess the operator's training level both in terms of task performance and requirement of cognitive activation, appears very useful. Such concern is based on the hypothesis that, during a training period, the execution of the task becomes more automatic, and thus less cognitive resources are required: consequently, a higher amount of cognitive resources will be available. In fact, after having been sufficiently practiced, the long-term memory will contain huge amount of domain-specific knowledge structures that can be described as hierarchically organized schemas that allow us to categorize different problems' states and decide the most appropriate solution moves. These schemas can operate under automatic—rather than controlled—processing (Kotovsky et al. 1985;

Schneider and Shiffrin 1977; Shiffrin and Schneider 1977). Automatic processing of schemas requires minimal working memory resources and allows problem-solving to proceed with minimal effort. In the *cognitive load theory* (CLT), it has been highlighted how the proper allocation of available cognitive resources is essential to learning (Sweller 1999; Sweller et al. 1998). The effects of practice at a cognitive level must be carefully considered in the interpretation of any study, as the changes in cognitive processes underlying task performance can be a key to understanding the changes in observed functional activations. Many researches have revealed that both developmental changes and changes in response to experience can occur at multiple levels of the *central nervous system* (CNS), from changes at the molecular or synaptic level, to changes in cortical maps and large-scale neural networks (Kelly and Garavan 2005). Across studies, three main patterns of practice-related activation change can be distinguished. Practice may result in an *increase or a decrease in activation* in the brain areas involved in the task performance, or it may produce a *functional reorganization* of brain activity, which is a combined pattern of activation increases and decreases across a number of brain areas (Petersen et al. 1998; Poldrack 2000; Raichle et al. 2001). *Increased neural efficiency* may correspond to a sharpening of the response in a particular neural network so that only a minority of neurons now fire strongly in response to a particular task or stimulus (Poldrack 2000). Practice-related activation decreases are therefore the result of a more efficient use of specific "neuronal circuits" (Petersen et al. 1998). In topographically organized cortex, this will be seen as a reduction in the spatial extent of activation, or, in areas with distributed representations, as a reduction in activation (Poldrack 2000). The term *increases in activation* refers to both practice-related expansions in cortical representations and increases in the strength of activations. Expanded representations may be reflected in increased strength of activations. On a neural level, increases in activation are suggested to reflect recruitment of additional cortical units with practice, seen in topographically organized cortex as an increase in the spatial extent of activation, or a strengthening of response within a region, observed as an increase in the level of activity within that region (Poldrack 2000). Practice-related reorganization of the functional anatomy of task performance may also be distinguished into two types: one constituting a *redistribution* and one constituting a *"true" reorganization*. The first might be considered as a pseudo-reorganization, or redistribution of functional activations. It constitutes a combination of increases and decreases in activation such that the task activation map generally contains the same areas at the end as at the beginning of practice, but the levels of activation within those areas have changed. The functional anatomy of the task therefore remains basically the same but the contribution of specific areas to task performance changes as a result of practice. Petersen et al. (1998) have discussed this topic in terms of a "scaffolding storage" framework, according to which a "scaffolding" set of regions is used to cope with novel demands during unskilled, effortful performance. After practice, processes or associations are more efficiently stored and accessed, and the scaffolding network falls away, evinced by decreased activation in those "scaffolding" attentional and control brain areas. Therefore, activations seen in previous practice involve generic attentional and control areas—

*prefrontal cortex* (PFC), *anterior cingulate cortex*, (ACC) *and posterior parietal cortex* (PPC) are the main areas considered to perform the "scaffolding" role, consistent with Sadmehr's and Holcomb's theories (Shadmehr and Holcomb 1997). The PFC, which has been previously implicated in goal-based and planning processes (Shallice 1982), is most readily associated with the "goal processor", whose primary function is to guide novice behavior by instantiating the task context (e.g., the instructions) and then structuring the sequential execution of task-relevant operations (Chein and Schneider 2005). Correspondingly, Cohen and colleagues (Cohen et al. 1996) have argued that the prefrontal cortex, in particular the dorsolateral part, serves to maintain goal-related information that biases subsequent processing according to task demands. By this account, practice-related decreases observed in the *dorsolateral prefrontal cortex* (DLPFC) may occur as a practice that strengthens the stimulus–response associations between paired items: therefore, the correct responses are potentiated, and the need for monitoring and biasing from task context (and thus the prefrontal cortex) is reduced. From a working memory perspective, the activity in these regions during the encoding phase may reflect material nonspecific rehearsal processes to maintain online the information, while it is being more permanently embedded into long-term memory (Buckner 1996).

The empirically observed domain-general network also included an extensive PPC where the activity decreased with practice. This region, which is widely viewed as an attentional center of the brain (Corbetta et al. 1995; Desimone 1992; Posner and Petersen 1990; Vandenberghe et al. 2012; Vandenberghe and Gillebert 2013), subserves a control process that allocates attention to task-specific information processing regions. Such attentional modulation is imposed by an "attention controller" that guides the information flow by regulating the output of the distributed information processing modules, during the novice's performance.

Practice also led to decreased activation of the ACC for both verbal and nonverbal paired associates: this region has a function corresponding to a sort of "activity monitor" in the control system. The activity monitor acts to track and records recent activity levels in task-relevant areas, as well as to establish a decision threshold for identifying strongly activated areas to the goal processor. Indeed, much of the recent literature on the ACC function has emphasized its role in communicating a need for greater cognitive control to other executive brain regions (Carter et al. 1999). Accordingly, this region's decrease in activation with practice may correspond to the decreased sampling time needed to select a decision, as familiarity with the correct pairings increased.

The second type of functional reorganization is considered to reflect a "*true*" *reorganization* of activations. It is observed as a change in the location of activations is associated with a shift in the cognitive processes underlying task performance. This shift ("process switching": see Dayan and Cohen 2011; Poldrack 2002) means that neurobiologically and cognitively, different tasks are being performed at the beginning and at the end of the practice, resulting in a coordinated increase and decrease of activation in separate brain regions, and it is consistent with studies demonstrating a reorganization of activation as a result of explicit shifts or differences in task strategy (e.g., Ashby et al. 2010; Bernstein et al. 2002; Doyon and

Benali 2005; Glabus et al. 2003; Hikosaka et al. 2002; Sakai et al. 1998). The results of these studies are also consistent with those of Seidler et al. (2002) who found out that, during the practice of a serial reaction time (SRT) task, there are learning-related activations in the premotor, motor, prefrontal, cingulate, and parietal cortices, as well as in the thalamus. Activation in these areas was observed to dissipate with practice even though a distractor task prevented the behavioral expression of learning. On the other hand, cerebellar activation increased during the expression of performance improvements, once the distractor had been removed. Hikosaka et al. (2002) have also demonstrated that automatic execution of a motor adaptation task produces long-term plastic changes in the *cortico-cerebellar* (CC) instead of the *cortico-striatal* (CS) system. Automatic execution was assessed by verifying that subjects had reached a complete asymptotic performance, and by testing such performance in a dual-task condition. Although activations in the *putamen* and other motor areas were observed in the early learning phase, increased activity in the *cerebellum* and parietal cortex were seen after the training, suggesting that the *cerebellum* and the associated cortical regions are sufficient to mediate automatic adapted movements. Both these studies demonstrated the "scaffolding" role of the frontal and parietal areas, activation which is subject to "pruning" as performance improves and becomes increasingly reliant on task-specific and performance-related areas located posteriorly in the brain. Furthermore, they demonstrated how learning-related changes in activation may occur in the absence of performance change, emphasizing the complexity of the brain's response to a repeated task experience.

As quoted above, the literature dealing with the effects of practice on the functional anatomy of task performance is extensive and complex, comprising a wide range of papers from disparate research perspectives. An interesting question then arises: is it possible to follow such changes of the cerebral activity along the training period, by using neuroimaging methodologies? This capability would provide additional and objective evidences related to the acquisition of ability during training phases, besides the overt behavioral measurements. The neuroimaging methodologies would be able to track reallocation of such cognitive resources during the training phases, providing an independent metric on the quality of the performed task. Neuroimaging methods such as *positron emission tomography* (PET) and *functional magnetic resonance imaging* (fMRI) are excellent tools in this endeavor, enabling the examination of how the brain adapts itself in response to practice or repeated exposure to particular tasks. However, their limitations in terms of cost, space, and invasiveness make them not suitable for real working environments, where less invasive approaches would be preferable and the costs for their implementations and usage have to be limited. In fact, PET and fMRI techniques require expensive devices, big customized rooms and high maintenance costs. In order to deal with these requirements and limitations, it has been investigated the possibility to track training progress of the operator by recording its brain activity by using the *electroencephalography* (EEG) technology, both in controlled environments (Labs) and in realistic operational settings.

## 3.5 Crew Resourses Management

Commercial and military aviation have been utilizing *crew resource management* (CRM) training for more than two decades (Kemper et al. 2016; Salas et al. 1999, 2001, 2006). CRM is an instructional strategy that trains crews to effectively use all of their available resources (i.e., people, equipment, and information; Helmreich et al. 1999). CRM training has been defined as a set of "*instructional strategies designed to improve teamwork in the cockpit by applying well tested training tools (*e.g., *performance measures, exercises, feedback mechanisms) and appropriate training methods (*e.g., *simulators, lectures, videos) targeted at specific content (*i.e., *teamwork knowledge, skills, and attitudes)*" (Salas et al. 1999). Despite its long history and touted success within aviation, some have begun to argue that the true impact of CRM training on aviation safety is still not understood. Salas et al. (2001) conducted a comprehensive review of studies published between 1983 and 1999 that evaluated the effectiveness of CRM training for flight crews within the commercial and military aviation domains. The review identified 58 studies suggesting that the data are encouraging, but the picture is not as clear as it should be after 20 years of implementation. In fact, a recent work indicated that the number of accidents involving a breakdown of CRM as a causal factor has remained fairly consistent over time despite the presence of CRM training. Specifically, Wiegmann and Shappell (2012) examined a breakdown of the factors contributing to U.S. Navy and Marine Corps accidents and found that in approximately 60% of the accidents between 1991 and 2000, a CRM failure in the cockpit was a factor. Similarly, an examination of aviation accidents during 1991 and 2000 indicated that approximately 41% of Part 121 (i.e., *aircraft conducting domestic commercial or flag operations*) and 23% of Part 135 (i.e., *aircraft conducting commuter or ondemand operations*) accidents involved a breakdown in CRM by the flight crew (Shappell and Wiegmann 2012). The full impact of CRM training on safety cannot yet be ascertained. Furthermore, a recent study by O'Connor et al. (2002) investigated the typical methods used to evaluate CRM training's effectiveness in the UK's aviation industry. Their research found that 69% evaluated training in terms of reactions, 21% measured learning in terms of attitudes, 36% measured learning in terms of knowledge tests, 53% measured behaviors, and 33% evaluated the training's impact on the organization. These numbers are relatively consistent with what is published in the literature. Despite this, it is interesting that other domains have begun to implement CRM training. For example, the medical community has implemented CRM training in the operating room, emergency room, and intensive care units. Other communities, such as nuclear power and offshore oil production, have also implemented CRM training for employees (Chan et al. 2016; Kemper et al. 2016). Therefore, the purpose of our studies is to investigate neurophysiological bases of cooperative behaviors for applications where the cooperation between two or more people is mandatory for the success of the task. This is the case, among many others, of commercial pilots whose operational works require the exchange of information and the synchronization of their activities in order to successfully execute the flight mission. Despite the importance of crew cooperation, during the last 30 years, most of the literature about the topic of aviation has described cerebral and neuroelectrical aspects associated with the activities and abilities of single pilots (Caldwell 2005; Dahlstrom and Nahlinder 2009), error management (Helmreich 2000), and prevention of accidents caused by pilots' involuntary sleep (Caldwell et al. 2009; Gundel et al. 1995). It is clear that the study of cerebral counterparts of efficient cooperation between pilots could open the way to a quantification of such "social interactions." The simultaneous recording of the cerebral activity from different subjects and the study of the synchronization between their brains were first proposed in fMRI as tools for the study of common vision (Hasson et al. 2004), and then translated to the neuroelectrical fields for the analysis of temporal motor synchronization (Cui et al. 2012; Hari et al. 2013; Konvalinka et al. 2014), shared attention (Bilek et al. 2015) and decision-making (Astolfi et al. 2011a, b; Tang et al. 2016). The introduction of the electroencephalogram (EEG) *hyperscanning* approach, as well as the portable EEG recording system, provided ecological experimental settings whereby people are able to interact naturally (Astolfi et al. 2011a, b; Dumas et al. 2010). This allowed to design experiments in more realistic environments (De Vico Fallani et al. 2010; Dumas et al. 2010; Toppi et al. 2016).

## References

Aguinis H, Kraiger K (2009) Benefits of training and development for individuals and teams, organizations, and society. Annu Rev Psychol 60:451–474. doi:10.1146/annurev.psych.60.110707.163505

Allport A (1988) What concept of consciousness? In: Marcel AJ, Bisiach E (eds) Consciousness in contemporary science. Oxford University Press

Anderson JR (1993) Rules of the mind. Psychology Press

Aricò P, Borghini G, Graziani I, Imbert J-P, Granger G, Benhacene R, Pozzi S, Napoletano L, Di Flumeri G, Colosimo A, Babiloni F (2015) Air-Traffic-Controllers (ATCO)—neurophysiological analysis of training and workload ATCO. Ital J Aerosp Med 12:18–35

Ashby FG, Turner BO, Horvitz JC (2010) Cortical and basal ganglia contributions to habit learning and automaticity. Trends Cogn Sci 14:208–215. doi:10.1016/j.tics.2010.02.001

Astolfi L, Toppi J, Borghini G, Vecchiato G, Isabella R, Fallani FDV, Cincotti F, Salinari S, Mattia D, He B, Caltagirone C, Babiloni F (2011a) Study of the functional hyperconnectivity between couples of pilots during flight simulation: an EEG hyperscanning study. In: Proceedings of the 2011 Annual International Conference of the IEEE Engineering in Medicine and Biology Society (EMBC), pp 2338–2341. doi:10.1109/IEMBS.2011.6090654

Astolfi Laura, Toppi J, De Vico Fallani F, Vecchiato G, Cincotti F, Wilke CT, Yuan H, Mattia D, Salinari S, He B, Babiloni F (2011b) Imaging the social brain by simultaneous hyperscanning during subject interaction. IEEE Intell Syst 26:38–45. doi:10.1109/MIS.2011.61

Ayaz H, Shewokis PA, Bunce S, Izzetoglu K, Willems B, Onaral B (2012) Optical brain monitoring for operator training and mental workload assessment. NeuroImage 59:36–47. doi:10.1016/j.neuroimage.2011.06.023

Ayaz H, Onaral B, Izzetoglu K, Shewokis PA, McKendrick R, Parasuraman R (2013) Continuous monitoring of brain dynamics with functional near infrared spectroscopy as a tool for neuroergonomic research: empirical examples and a technological development. Front Hum Neurosci 7. doi:10.3389/fnhum.2013.00871

Baars BJ (1997) Some essential differences between consciousness and attention, perception, and working memory. Conscious Cogn 6:363–371

Baldwin CL (2003) Commentary. Theor Issues Ergon Sci 4:132–141. doi:10.1080/1463922 0210159807

Barach P, Small SD (2000) Reporting and preventing medical mishaps: lessons from non-medical near miss reporting systems. BMJ 320:759–763

Barach P, Weingart M (2004) Trauma team performance. In: Trauma: resuscitation, anesthesia, surgery, and critical care. Dekker, Inc., New York

Bellenkes CAH (2007) Contemporary issues in human factors and aviation safety. Ergonomics 50:963–965. doi:10.1080/00140130600971093

Berka C, Levendowski DJ, Cvetinovic MM, Petrovic MM, Davis G, Lumicao MN, Zivkovic VT, Popovic MV, Olmstead R (2004) Real-time analysis of EEG indexes of alertness, cognition, and memory acquired with a wireless EEG headset. Int J Hum Comput Interact 17:151–170. doi:10.1207/s15327590ijhc1702_3

Bernard C (1957) An introduction to the study of experimental medicine. Courier Corporation

Bernstein LJ, Beig S, Siegenthaler AL, Grady CL (2002) The effect of encoding strategy on the neural correlates of memory for faces. Neuropsychologia 40:86–98

Bert P (1878) La pression barométrique. G. Masson, Paris

Bherer L, Kramer AF, Peterson MS, Colcombe S, Erickson K, Becic E (2005) Training effects on dual-task performance: are there age-related differences in plasticity of attentional control? Psychol Aging 20:695–709. doi:10.1037/0882-7974.20.4.695

Bilek E, Ruf M, Schäfer A, Akdeniz C, Calhoun VD, Schmahl C, Demanuele C, Tost H, Kirsch P, Meyer-Lindenberg A (2015) Information flow between interacting human brains: identification, validation, and relationship to social expertise. Proc Natl Acad Sci 112:5207–5212. doi:10.1073/pnas.1421831112

Borghini G, Astolfi L, Vecchiato G, Mattia D, Babiloni F (2012) Measuring neurophysiological signals in aircraft pilots and car drivers for the assessment of mental workload, fatigue and drowsiness. Neurosci Biobehav Rev. doi:10.1016/j.neubiorev.2012.10.003

Borghini G, Arico P, Astolfi L, Toppi J, Cincotti F, Mattia D, Cherubino P, Vecchiato G, Maglione AG, Graziani I, Babiloni F (2013) Frontal EEG theta changes assess the training improvements of novices in flight simulation tasks. In: Proceedings of the 2013 35th Annual International Conference of the IEEE Engineering in Medicine and Biology Society (EMBC), pp 6619–6622. doi:10.1109/EMBC.2013.6611073

Borghini G, Arico P, Ferri F, Graziani I, Pozzi S, Napoletano L, Imbert JP, Granger G, Benhacene R, Babiloni F, (2014a) A neurophysiological training evaluation metric for air traffic management. In: Proceedings of the 2014 36th Annual International Conference of the IEEE Engineering in Medicine and Biology Society (EMBC), pp 3005–3008. doi:10.1109/EMBC.2014.6944255

Borghini G, Astolfi L, Vecchiato G, Mattia D, Babiloni F (2014b) Measuring neurophysiological signals in aircraft pilots and car drivers for the assessment of mental workload, fatigue and drowsiness. Neurosci Biobehav Rev 44:58–75. doi:10.1016/j.neubiorev.2012.10.003

Borghini G, Aricò P, Di Flumeri G, Salinari S, Colosimo A, Bonelli S, Napoletano L, Ferreira A, Babiloni F (2015) Avionic technology testing by using a cognitive neurometric index: a study with professional helicopter pilots. In: Proceedings of the 2015 37th Annual International Conference of the IEEE Engineering in Medicine and Biology Society (EMBC), pp 6182–6185. doi:10.1109/EMBC.2015.7319804

Broach D (2013) Selection of the next generation of air traffic control specialists: aptitude requirements for the air traffic control tower cab in 2018 (Aerospace Medicine Technical Reports No. DOT/FAA/AM-13/5). FAA Office of Aerospace Medicine Civil Aerospace Medical Institute

Broach D, Manning CA, United States. Office of Aviation Medicine (1997) Review of air traffic controller selection [microform]: an international perspective/Dana Broach, Carol A. Manning. U.S. Dept. of Transportation, Federal Aviation Administration, Office of Aviation Medicine; Available through the National Technical Information Service, Washington, D.C.: Springfield

Brookings JB, Wilson GF, Swain CR (1996) Psychophysiological responses to changes in workload during simulated air traffic control. Biol Psychol 42:361–377

Buckner RL (1996) Beyond HERA: contributions of specific prefrontal brain areas to long-term memory retrieval. Psychon Bull Rev 3:149–158. doi:10.3758/BF03212413

Bunce SC, Izzetoglu K, Ayaz H, Shewokis P, Izzetoglu M, Pourrezaei K, Onaral B (2011) Implementation of fNIRS for monitoring levels of expertise and mental workload. In: Schmorrow DD, Fidopiastis CM (eds) Foundations of augmented cognition. Directing the future of adaptive systems. Springer, Berlin, pp 13–22

Cabeza R, Nyberg L (2000) Imaging cognition II: an empirical review of 275 PET and fMRI studies. J Cogn Neurosci 12:1–47

Cain B (2007) A review of the mental workload literature

Caldwell JA (2005) Fatigue in aviation. Travel Med Infect Dis 3:85–96. doi:10.1016/j.tmaid.2004.07.008

Caldwell JA, Mallis MM, Caldwell JL, Paul MA, Miller JC, Neri DF (2009) Fatigue countermeasures in aviation. Aviat Space Environ Med 80:29–59. doi:10.3357/ASEM.2435.2009

Campbell JP (1971) Personnel training and development. Annu Rev Psychol 22:565–602. doi:10.1146/annurev.ps.22.020171.003025

Carter CS, Botvinick MM, Cohen JD (1999) The contribution of the anterior cingulate cortex to executive processes in cognition. Rev Neurosci 10:49–57

Castor M, Hanson E, Svensson E, Nählinder S, LeBlaye P, MacLeod I, Wright N, Alfredson J, Agren L, Berggren P (2003) GARTEUR Handbook of mental workload measurement (Technical chapter 145). Group for Aeronautical Research and Technology in Europe, Flight Mechanics Action Group FM AG13

Chan CKW, So H, Ng W, Chan P, Ma W, Chan K, Leung S, Ho L-Y (2016) Does classroom-based crew resource management training have an effect on attitudes between doctors and nurses? Int J Med Educ 7:109–114. doi:10.5116/ijme.56f5.6804

Chein JM, Schneider W (2005) Neuroimaging studies of practice-related change: fMRI and meta-analytic evidence of a domain-general control network for learning. Brain Res Cogn Brain Res 25:607–623. doi:10.1016/j.cogbrainres.2005.08.013

Cohen JD, Braver TS, O'Reilly RC (1996) A computational approach to prefrontal cortex, cognitive control and schizophrenia: recent developments and current challenges. Philos Trans R Soc Lond B Biol Sci 351:1515–1527. doi:10.1098/rstb.1996.0138

Corbetta M, Shulman GL (2002) Control of goal-directed and stimulus-driven attention in the brain. Nat Rev Neurosci 3:201–215. doi:10.1038/nrn755

Corbetta M, Shulman GL, Miezin FM, Petersen SE (1995) Superior parietal cortex activation during spatial attention shifts and visual feature conjunction. Science 270:802–805

Costa G (1993) Evaluation of workload in air traffic controllers. Ergonomics 36:1111–1120. doi:10.1080/00140139308967982

Cui X, Bray S, Bryant DM, Glover GH, Reiss AL (2011) A quantitative comparison of NIRS and fMRI across multiple cognitive tasks. NeuroImage 54:2808–2821. doi:10.1016/j.neuroimage.2010.10.069

Cui X, Bryant DM, Reiss AL (2012) NIRS-based hyperscanning reveals increased interpersonal coherence in superior frontal cortex during cooperation. NeuroImage 59:2430–2437. doi:10.1016/j.neuroimage.2011.09.003

Dahlstrom N, Nahlinder S (2009) Mental workload in aircraft and simulator during basic civil aviation training. Int J Aviat Psychol 19:309–325. doi:10.1080/10508410903187547

Dayan E, Cohen LG (2011) Neuroplasticity subserving motor skill learning. Neuron 72:443–454. doi:10.1016/j.neuron.2011.10.008

De Vico Fallani F, Nicosia V, Sinatra R, Astolfi L, Cincotti F, Mattia D, Wilke C, Doud A, Latora V, He B, Babiloni F (2010) Defecting or not defecting: how to "read" human behavior during cooperative games by EEG measurements. PLoS ONE 5:e14187. doi:10.1371/journal.pone.0014187
De Waard (1996) The measurement of drivers' mental workload. University of Groningen, Traffic Research Centre, Netherlands
Dehais F, Tessier C, Christophe L, Reuzeau F (2010) The perseveration syndrome in the pilot's activity: guidelines and cognitive countermeasures. In: Palanque P, Vanderdonckt J, Winckler M (eds) Human error, safety and systems development. Lecture notes in computer science. Springer, Berlin, pp 68–80
Derosière G, Mandrick K, Dray G, Ward TE, Perrey S (2013) NIRS-measured prefrontal cortex activity in neuroergonomics: strengths and weaknesses. Front Hum Neurosci 7. doi:10.3389/fnhum.2013.00583
Desimone R (1992) Neural circuits for visual attention in the primate brain. In: Carpenter GA, Grossberg S (eds) Neural networks for vision and image processing. The MIT Press, Cambridge, pp 343–364
Di Flumeri G, Borghini G, Aricò P, Pozzi S, Bonelli S, Golfetti A, Colosimo A, Babiloni F (2015) Mental workload evaluation of ATCOs during ecological ATM scenarios. Ital J Aerosp, Med 1
Doyon J, Benali H (2005) Reorganization and plasticity in the adult brain during learning of motor skills. Curr Opin Neurobiol 15:161–167. doi:10.1016/j.conb.2005.03.004
Dumas G, Nadel J, Soussignan R, Martinerie J, Garnero L (2010) Inter-brain synchronization during social interaction. PLoS ONE 5:e12166. doi:10.1371/journal.pone.0012166
Durantin G, Gagnon J-F, Tremblay S, Dehais F (2014) Using near infrared spectroscopy and heart rate variability to detect mental overload. Behav Brain Res 259:16–23. doi:10.1016/j.bbr.2013.10.042
Duru DG, Duru AD, Barkana DE, Sanli O, Ozkan M (2013) Assessment of surgeon's stress level and alertness using EEG during laparoscopic simple nephrectomy. In: Proceedings of the 2013 6th International IEEE/EMBS Conference on Neural Engineering (NER), pp 452–455. doi:10.1109/NER.2013.6695969
East JA (2000) Feature selection for predicting pilot mental workload
Eggemeier FT, Wilson GF, Kramer AF, Damos DL (1991) Workload assessment in multi-task environments. Multiple-task performance. Taylor & Francis, London, pp 207–216
Erickson KI, Colcombe SJ, Wadhwa R, Bherer L, Peterson MS, Scalf PE, Kim JS, Alvarado M, Kramer AF (2007) Training-induced functional activation changes in dual-task processing: an FMRI study. Cereb. Cortex N Y N 1991 17:192–204. doi:10.1093/cercor/bhj137
Fairclough SH, Venables L, Tattersall A (2005) The influence of task demand and learning on the psychophysiological response. Int J Psychophysiol Off J Int Organ Psychophysiol 56:171–184. doi:10.1016/j.ijpsycho.2004.11.003
Fitts PM, Posner MI (1979) Human performance. Greenwood Press, Westport
Gevins A, Smith ME (2003) Neurophysiological measures of cognitive workload during human-computer interaction. Theor Issues Ergon, Sci 4
Gevins A, Smith ME, McEvoy L, Yu D (1997) High-resolution EEG mapping of cortical activation related to working memory: effects of task difficulty, type of processing, and practice. Cereb Cortex N Y N 1991 7:374–385
Glabus MF, Horwitz B, Holt JL, Kohn PD, Gerton BK, Callicott JH, Meyer-Lindenberg A, Berman KF (2003) Interindividual differences in functional interactions among prefrontal, parietal and parahippocampal regions during working memory. Cereb Cortex N Y N 1991 13:1352–1361
Glass JM, Schumacher EH, Lauber EJ, Zurbriggen EL, Gmeindl L, Kieras DE, Meyer DE (2000) Aging and the psychological refractory period: task-coordination strategies in young and old adults. Psychol Aging 15:571–595. doi:10.1037//0882-7974.15.4.571
Goldberg DH, Vogelstein RJ, Socolinsky DA, Wolff LB (2011) Toward a wearable, neurally-enhanced augmented reality system. In: Schmorrow DD, Fidopiastis CM (eds) Foundations of augmented cognition. Directing the future of adaptive systems. Springer, Berlin, pp 493–499

Goldstein IL (1980) Training in work organizations. Annu Rev Psychol 31:229–272. doi:10.1146/annurev.ps.31.020180.001305
Goldstein IL, Ford JK (2002) Training in organizations: needs assessment, development, and evaluation. Wadsworth
Gopher D (1993) The skill of attention control: acquisition and execution of attention strategies. In: Meyer DE, Kornblum S (eds) Attention and performance XIV (Silver Jubilee Volume). MIT Press, Cambridge, pp 299–322
Gopher D, Donchin E (1986) Workload: an examination of the concept. In: Boff KR, Kaufman L, Thomas JP (eds) Handbook of perception and human performance. Cognitive processes and performance, vol 2. Wiley, Oxford, pp 1–49
Gundel A, Drescher J, Maaß H, Samel A, Vejvoda M (1995) Sleepiness of civil airline pilots during two consecutive night flights of extended duration. Biol Psychol, EEG in Basic and Applied Settings 40:131–141. doi:10.1016/0301-0511(95)05107-4
Hamman W (2004) The complexity of team training: what we have learned from aviation and its applications to medicine. Qual Saf Health Care 13:i72–i79. doi:10.1136/qshc.2004.009910
Hankins TC, Wilson GF (1998) A comparison of heart rate, eye activity, EEG and subjective measures of pilot mental workload during flight. Aviat Space Environ Med 69:360–367
Hari R, Himberg T, Nummenmaa L, Hämäläinen M, Parkkonen L (2013) Synchrony of brains and bodies during implicit interpersonal interaction. Trends Cogn Sci 17:105–106. doi:10.1016/j.tics.2013.01.003
Harrison J, Izzetoglu K, Ayaz H, Willems B, Hah S, Ahlstrom U, Woo H, Shewokis PA, Bunce SC, Onaral B (2014) Cognitive workload and learning assessment during the implementation of a next-generation air traffic control technology using functional near-infrared spectroscopy. IEEE Trans Hum Mach Syst 44:429–440. doi:10.1109/THMS.2014.2319822
Hart SG, Staveland LE (1988) Development of NASA-TLX (Task Load Index): results of empirical and theoretical research. In: Human mental workload. North-Holland, pp 139–183
Hart SG, Wickens CD (1990) Workload assessment and prediction. In: Booher HR (ed) Manprint. Springer, Netherlands, pp 257–296
Hasson U, Nir Y, Levy I, Fuhrmann G, Malach R (2004) Intersubject synchronization of cortical activity during natural vision. Science 303:1634–1640. doi:10.1126/science.1089506
Helmreich RL (2000) On error management: lessons from aviation. BMJ 320:781–785
Helmreich RL, Merritt AC, Wilhelm JA (1999) The evolution of crew resource management training in commercial aviation. Int J Aviat Psychol 9:19–32. doi:10.1207/s15327108ijap0901_2
Herff C, Heger D, Fortmann O, Hennrich J, Putze F, Schultz T (2013) Mental workload during n-back task-quantified in the prefrontal cortex using fNIRS. Front Hum Neurosci 7:935. doi:10.3389/fnhum.2013.00935
Hikosaka O, Nakamura K, Sakai K, Nakahara H (2002) Central mechanisms of motor skill learning. Curr Opin Neurobiol 12:217–222. doi:10.1016/S0959-4388(02)00307-0
Hill C, Lent R (2006) A narrative and meta-analytic review of helping skills training: time to revive a dormant area of inquiry. Psychother Theory Res Pract 43:154–172
Huppert TJ, Hoge RD, Diamond SG, Franceschini MA, Boas DA (2006) A temporal comparison of BOLD, ASL, and NIRS hemodynamic responses to motor stimuli in adult humans. NeuroImage 29:368–382. doi:10.1016/j.neuroimage.2005.08.065
Izzetoglu K, Bunce S, Onaral B, Pourrezaei K, Chance B (2004) Functional optical brain imaging using near-infrared during cognitive tasks. Int J Hum Comput Interact 17:211–227. doi:10.1207/s15327590ijhc1702_6
Izzetoglu K, Ayaz H, Hing JT, Shewokis PA, Bunce SC, Oh P, Onaral B (2015) UAV operators workload assessment by optical brain imaging technology (fNIR). In: Valavanis KP, Vachtsevanos GJ (eds) Handbook of unmanned aerial vehicles. Springer, Netherlands, pp 2475–2500. doi:10.1007/978-90-481-9707-1_22
Jaušovec N, Jaušovec K (2012) Working memory training: improving intelligence–changing brain activity. Brain Cogn 79:96–106. doi:10.1016/j.bandc.2012.02.007

Jorna PG (1992) Spectral analysis of heart rate and psychological state: a review of its validity as a workload index. Biol Psychol 34:237–257

Jorna PG (1993) Heart rate and workload variations in actual and simulated flight. Ergonomics 36:1043–1054. doi:10.1080/00140139308967976

Kellogg RT (1980) Is conscious attention necessary for long-term storage? J Exp Psychol [Hum Learn] 6:379–390. doi:10.1037/0278-7393.6.4.379

Kelly AMC, Garavan H (2005) Human functional neuroimaging of brain changes associated with practice. Cereb Cortex N Y N 1991 15:1089–1102. doi:10.1093/cercor/bhi005

Kemper PF, de Bruijne M, van Dyck C, So RL, Tangkau P, Wagner C (2016) Crew resource management training in the intensive care unit. A multisite controlled before–after study. BMJ Qual Saf 25:577–587. doi:10.1136/bmjqs-2015-003994

Klimesch W (1999) EEG alpha and theta oscillations reflect cognitive and memory performance: a review and analysis. Brain Res Rev 29:169–195. doi:10.1016/S0165-0173(98)00056-3

Klimesch W, Doppelmayr M, Pachinger T, Ripper B (1997) Brain oscillations and human memory: EEG correlates in the upper alpha and theta band. Neurosci Lett 238:9–12

Kochanska G, Murray KT, Harlan ET (2000) Effortful control in early childhood: continuity and change, antecedents, and implications for social development. Dev Psychol 36:220–232

Konvalinka I, Bauer M, Stahlhut C, Hansen LK, Roepstorff A, Frith CD (2014) Frontal alpha oscillations distinguish leaders from followers: multivariate decoding of mutually interacting brains. NeuroImage 94:79–88. doi:10.1016/j.neuroimage.2014.03.003

Kotovsky K, Hayes JR, Simon HA (1985) Why are some problems hard? Evidence from Tower of Hanoi. Cogn Psychol 17:248–294. doi:10.1016/0010-0285(85)90009-X

Kozlowski SWJ, Gully SM, Brown KG, Salas E, Smith EM, Nason ER (2001) Effects of training goals and goal orientation traits on multidimensional training outcomes and performance adaptability. Organ Behav Hum Decis Process 85:1–31. doi:10.1006/obhd.2000.2930

Kramer AF, Larish JF, Strayer DL (1995) Training for attentional control in dual task settings: a comparison of young and old adults. J Exp Psychol Appl 1:50–76. doi:10.1037/1076-898X.1.1.50

Kramer AF, Larish JL, Weber TA, Bardell L (1999) Training for executive control: task coordination strategies and aging. In: Gopher D, Koriat A (eds) Attention and performance XVII: Cognitive Regulation of performance: interaction of theory and application, attention and performance. The MIT Press, Cambridge, pp 617–652

LaBerge D (1995) Attentional processing: the brain's art of mindfulness. Harvard University Press

Leape LL, Fromson JA (2006) Problem doctors: is there a system-level solution? Ann Intern Med 144:107–115

Lee DH, Park KS (1990) Multivariate analysis of mental and physical load components in sinus arrhythmia scores. Ergonomics 33:35–47. doi:10.1080/00140139008927092

Leonard M, Tarrant C (2001) Culture, systems, and human factors—two tales of patient safety: the KP Colorado region's experience. Permanente J 5:46–49

Leonard M, Graham S, Bonacum D (2004) The human factor: the critical importance of effective teamwork and communication in providing safe care. Qual Saf Health Care 13:i85–i90. doi:10.1136/qshc.2004.010033

Liao L-D, Lin C-T, McDowell K, Wickenden AE, Gramann K, Jung T-P, Ko L-W, Chang J-Y (2012) Biosensor technologies for augmented brain-computer interfaces in the next decades. Proc IEEE 100:1553–1566. doi:10.1109/JPROC.2012.2184829

Logan GD (1988) Toward an instance theory of automatization. Psychol Rev 95:492–527

Marcel AJ (1983) Conscious and unconscious perception: experiments on visual masking and word recognition. Cogn Psychol 15:197–237. doi:10.1016/0010-0285(83)90009-9

Medvedev AV, Kainerstorfer J, Borisov SV, Barbour RL, VanMeter J (2008) Event-related fast optical signal in a rapid object recognition task: improving detection by the independent component analysis. Brain Res 1236:145–158. doi:10.1016/j.brainres.2008.07.122

Miller S (2001) Literature review: workload measures, Document N01-006. National Advanced Driving Simulator

Models of Working Memory [WWW Document] (n.d.) Cambridge University Press. URL http://www.cambridge.org/it/academic/subjects/psychology/cognition/models-working-memory-mechanisms-active-maintenance-and-executive-control?format=PB&isbn=9780521587211 (Accessed 3.4.17)

Müller K-R, Tangermann M, Dornhege G, Krauledat M, Curio G, Blankertz B (2008) Machine learning for real-time single-trial EEG-analysis: from brain-computer interfacing to mental state monitoring. J Neurosci Methods 167:82–90. doi:10.1016/j.jneumeth.2007.09.022

Murkin JM, Arango M (2009) Near-infrared spectroscopy as an index of brain and tissue oxygenation. Br J Anaesth 103(Suppl 1):i3–13. doi:10.1093/bja/aep299

Neumann O (1984) Automatic processing: a review of recent findings and a plea for an old theory. In: Prinz W, Sanders AF (eds) Cognition and motor processes. Springer, Berlin, pp 255–293

Newell A, Rosenbloom PS (1993) The Soar papers, vol 1. In: Rosenbloom PS, Laird JE, Newell A (eds). MIT Press, Cambridge, pp 81–135

Norman DA, Shallice T (1986) Attention to action. In: Davidson RJ, Schwartz GE, Shapiro D (eds) Consciousness and self-regulation. Springer, Boston, pp 1–18

O'Connor P, Hörmann H-J, Flin R, Lodge M, Goeters K-M, Group TJ (2002) Developing a method for evaluating crew resource management skills: a European perspective. Int J Aviat Psychol 12:263–285. doi:10.1207/S15327108IJAP1203_5

Onton J, Delorme A, Makeig S (2005) Frontal midline EEG dynamics during working memory. NeuroImage 27:341–356. doi:10.1016/j.neuroimage.2005.04.014

Owen AM, McMillan KM, Laird AR, Bullmore E (2005) N-back working memory paradigm: a meta-analysis of normative functional neuroimaging studies. Hum Brain Mapp 25:46–59. doi:10.1002/hbm.20131

Parasuraman R, Hancock P (2001) Adaptive control of mental workload. In: Stress, workload, and fatigue. pp 305–320

Parasuraman R, McKinley RA (2014) Using noninvasive brain stimulation to accelerate learning and enhance human performance. J Hum Factors Ergon Soc. 0018720814538815. doi:10.1177/0018720814538815

Petersen SE, van Mier H, Fiez JA, Raichle ME (1998) The effects of practice on the functional anatomy of task performance. Proc Natl Acad Sci 95:853–860

Plichta MM, Herrmann MJ, Baehne CG, Ehlis A-C, Richter MM, Pauli P, Fallgatter AJ (2006) Event-related functional near-infrared spectroscopy (fNIRS): are the measurements reliable? NeuroImage 31:116–124. doi:10.1016/j.neuroimage.2005.12.008

Polanyi M (2012) Personal knowledge. Routledge

Poldrack RA (2000) Imaging brain plasticity: conceptual and methodological issues–a theoretical review. NeuroImage 12:1–13. doi:10.1006/nimg.2000.0596

Poldrack RA (2002) Neural systems for perceptual skill learning. Behav Cogn Neurosci Rev 1: 76–83. doi:10.1177/1534582302001001005

Polich J (1986) Attention, probability, and task demands as determinants of P300 latency from auditory stimuli. Electroencephalogr Clin Neurophysiol 63:251–259

Posner MI, Petersen SE (1990) The attention system of the human brain. Annu Rev Neurosci 13:25–42. doi:10.1146/annurev.ne.13.030190.000325

Posner MI, Rothbart MK (2007) Research on attention networks as a model for the integration of psychological science. Annu Rev Psychol 58:1–23. doi:10.1146/annurev.psych.58.110405.085516

Raichle ME, MacLeod AM, Snyder AZ, Powers WJ, Gusnard DA, Shulman GL (2001) A default mode of brain function. Proc Natl Acad Sci 98:676–682. doi:10.1073/pnas.98.2.676

Rasmussen J (1983) Skills, rules, and knowledge; signals, signs, and symbols, and other distinctions in human performance models. IEEE Trans Syst Man Cybern SMC-13, 257–266. doi:10.1109/TSMC.1983.6313160

Rasmussen J (1986) Information processing and human-machine interaction: an approach to cognitive engineering. North-Holland

Rasmussen J, Vicente K (1989) Coping with human errors through system design: implications for ecological interface design. Int J Man Mach Stud 31:517–534. doi:10.1016/0020-7373(89)90014-X

Raz A, Buhle J (2006) Typologies of attentional networks. Nat Rev Neurosci 7:367–379. doi:10.1038/nrn1903

Reid GB, Nygren TE (1988) The subjective workload assessment technique: a scaling procedure for measuring mental workload. In: Meshkati N, Hancock PA (eds) Advances in psychology, Human mental workload. North-Holland, pp 185–218

Ritter F, Schooler L (2001) The learning curve. In: International encyclopedia of the social & behavioral sciences. Elsevier Science Ltd.

Rodgers MD, Drechsler GK (1993) Conversion of the CTA, Inc., en route operations concept database into a formal sentence outline job task taxonomy

Roscoe AH (1992) Assessing pilot workload. Why measure heart rate, HRV and respiration? Biol Psychol 34:259–287

Roscoe AH (1993) Heart rate as a psychophysiological measure for in-flight workload assessment. Ergonomics 36:1055–1062. doi:10.1080/00140139308967977

Rosenblueth A, Wiener N, Bigelow J (1943) Behavior, purpose and teleology. Philos Sci 10:18–24

Roy RN, Bonnet S, Charbonnier S, Campagne A (2016) Efficient workload classification based on ignored auditory probes: a proof of concept. Front Hum Neurosci 10. doi:10.3389/fnhum.2016.00519

Rubio S, Díaz E, Martín J, Puente JM (2004) Evaluation of subjective mental workload: a comparison of SWAT, NASA-TLX, and workload profile methods. Appl Psychol 53:61–86. doi:10.1111/j.1464-0597.2004.00161.x

Rudner M, Lunner T (2014) Cognitive spare capacity and speech communication: a narrative overview. Biomed Res Int 2014:869726. doi:10.1155/2014/869726

Ryu K, Myung R (2005) Evaluation of mental workload with a combined measure based on physiological indices during a dual task of tracking and mental arithmetic. Int J Ind Ergon 35:991–1009. doi:10.1016/j.ergon.2005.04.005

Sadaghiani S, Scheeringa R, Lehongre K, Morillon B, Giraud A-L, Kleinschmidt A (2010) Intrinsic connectivity networks, alpha oscillations, and tonic alertness: a simultaneous electroencephalography/functional magnetic resonance imaging study. J Neurosci Off J Soc Neurosci 30:10243–10250. doi:10.1523/JNEUROSCI.1004-10.2010

Sakai K, Hikosaka O, Miyauchi S, Takino R, Sasaki Y, Pütz B (1998) Transition of brain activation from frontal to parietal areas in visuomotor sequence learning. J Neurosci 18:1827–1840

Salas E, Prince C, Bowers CA, Stout RJ, Oser RL, Cannon-Bowers JA (1999) A methodology for enhancing crew resource management training. Hum Factors 41:161–172. doi:10.1518/001872099779577255

Salas Eduardo, Burke CS, Bowers CA, Wilson KA (2001) Team training in the skies: does Crew Resource Management (CRM) training work? Hum Factors 43:641–674. doi:10.1518/001872001775870386

Salas E, Wilson KA, Burke CS, Wightman DC (2006) Does crew resource management training work? An update, an extension, and some critical needs. Hum Factors 48:392–412. doi:10.1518/001872006777724444

Satterfield JM, Hughes E (2007) Emotion skills training for medical students: a systematic review. Med Educ 41:935–941. doi:10.1111/j.1365-2923.2007.02835.x

Schmidt RA, Lange C, Young DE (1990) Optimizing summary knowledge of results for skill learning. Hum Mov Sci 9:325–348. doi:10.1016/0167-9457(90)90007-Z

Schneider W, Shiffrin RM (1977) Controlled and automatic human information processing: I. Detection, search, and attention. Psychol Rev 84:1–66. doi:10.1037/0033-295X.84.1.1

Schumacher EH, Lauber EJ, Glass JM, Zurbriggen EL, Gmeindl L, Kieras DE, Meyer DE (1999) Concurrent response-selection processes in dual-task performance: evidence for adaptive executive control of task scheduling. J Exp Psychol Hum Percept Perform 25:791–814. doi:10.1037/0096-1523.25.3.791

Seidler RD, Purushotham A, Kim S-G, Uğurbil K, Willingham D, Ashe J (2002) Cerebellum activation associated with performance change but not motor learning. Science 296: 2043–2046. doi:10.1126/science.1068524

Selection of the next generation of air traffic control specialists: aptitude requirements for the air traffic control tower cab in PDF [WWW Document], (n.d.) URL http://docplayer.net/26521424-Selection-of-the-next-generation-of-air-traffic-control-specialists-aptitude-requirements-for-the-air-traffic-control-tower-cab-in-2018.html (Accessed 2.13.17)

Shadmehr R, Holcomb HH (1997) Neural correlates of motor memory consolidation. Science 277:821–825

Shallice T (1982) Specific impairments of planning. Philos Trans R Soc Lond B Biol Sci 298: 199–209

Shappell PSA, Wiegmann PDA (2012) A human error approach to aviation accident analysis: the human factors analysis and classification system. Ashgate Publishing, Ltd.

Shiffrin RM, Schneider W (1977) Controlled and automatic human information processing: II. Perceptual learning, automatic attending and a general theory. Psychol Rev 127–190

Shou G, Ding L, Dasari D (2012) Probing neural activations from continuous EEG in a real-world task: time-frequency independent component analysis. J Neurosci Methods 209:22–34. doi:10.1016/j.jneumeth.2012.05.022

Sirevaag EJ, Kramer AF, Wickens CD, Reisweber M, Strayer DL, Grenell JF (1993) Assessment of pilot performance and mental workload in rotary wing aircraft. Ergonomics 36:1121–1140. doi:10.1080/00140139308967983

Smith ME, Gevins A (2005) Neurophysiologic monitoring of mental workload and fatigue during operation of a flight simulator. pp 116–126. doi:10.1117/12.602181

Spearman C (1928) The origin of error. J Gen Psychol 1:29–53. doi:10.1080/00221309.1928.9923410

Stern JA, Skelly JJ (1984) The eye blink and workload considerations. In: Proceedings of the Human Factors Society—28th Annual Meeting. San Antonio, Texas

Sweller J (1999) Instructional design in technical areas. Australian Council for Educational

Sweller J, van Merrienboer JJG, Paas FGWC (1998) Cognitive architecture and instructional design. Educ Psychol Rev 10:251–296. doi:10.1023/A:1022193728205

Taggart WR (1994) Crew resource management: achieving enhanced flight operations. Aviat Psychol Pract 309–339

Tang R, Razi A, Friston KJ, Tang Y-Y (2016) Mapping smoking addiction using effective connectivity analysis. Front Hum Neurosci 10. doi:10.3389/fnhum.2016.00195

Tannenbaum S, Yukl G (1992) Training and development in work organizations. Annu Rev Psychol 43:399–441. doi:10.1146/annurev.ps.43.020192.002151

Tattersall AJ, Foord PS (1996) An experimental evaluation of instantaneous self-assessment as a measure of workload. Ergonomics 39:740–748

Toppi J, Borghini G, Petti M, He EJ, Giusti VD, He B, Astolfi L, Babiloni F (2016) Investigating cooperative behavior in ecological settings: an EEG hyperscanning study. PLoS ONE 11: e0154236. doi:10.1371/journal.pone.0154236

Van Orden KF, Makeig S, Jung TP, Limbert W (1999) Eye activity correlates of workload during a visuospatial memory task

Vandenberghe R, Gillebert CR (2013) Dissociations between spatial-attentional processes within parietal cortex: insights from hybrid spatial cueing and change detection paradigms. Front Hum Neurosci 7. doi:10.3389/fnhum.2013.00366

Vandenberghe R, Molenberghs P, Gillebert CR (2012) Spatial attention deficits in humans: the critical role of superior compared to inferior parietal lesions. Neuropsychologia, Special Issue: Spatial Neglect and Attention 50:1092–1103. doi:10.1016/j.neuropsychologia.2011.12.016

Veltman JA, Gaillard AW (1996) Physiological indices of workload in a simulated flight task. Biol Psychol 42:323–342
Venables L, Fairclough SH (2009) The influence of performance feedback on goal-setting and mental effort regulation. Motiv Emot 33:63–74. doi:10.1007/s11031-008-9116-y
Warm JS, Parasuraman R, Matthews G (2008) Vigilance requires hard mental work and is stressful. Hum Factors 50:433–441
Wexley KN (1984) Personnel training. Annu Rev Psychol 35:519–551. doi:10.1146/annurev.ps.35.020184.002511
Wickens CD (1984) Processing resources in attention. In: Varieties of attention. pp 62–102
Wickens CD, Hollands JG, Banbury S, Parasuraman R (2012) Engineering psychology & human performance, 4th edn. Psychology Press, Boston
Wilson GF (1993) Air-to-ground training missions: a psychophysiological workload analysis. Ergonomics 36:1071–1087. doi:10.1080/00140139308967979
Wilson GF, Fullenkamp P, Davis I (1994) Evoked potential, cardiac, blink, and respiration measures of pilot workload in air-to-ground missions. Aviat Space Environ Med 65:100–105
Wilson GF, Swain CR, Ullsperger P (1999) EEG power changes during a multiple level memory retention task. Int J Psychophysiol Off J Int Organ Psychophysiol 32:107–118

# Chapter 4
# Cognitive Processes

In the fourth chapter, detailed descriptions and scientific evidences regarding the cognitive processes mainly involved with the mental workload, cognitive control behavior, and training are provided. The use of neuroimaging techniques allows the analysis of cognitive processes underlying cognition and human behavior. A precise identification of neural responses across specific brain regions can provide information by which defining metric for the objective evaluation of the considered cognitive phenomenon. Among the main cognitive processes, the working memory (WM), attention, and decision-making can be considered as elementary parts for the study of the learning and the mental workload (Harris 2015). Throughout the chapter, the reader will understand why and how these elementary processes play key roles in the definition of the considered mental states. Therefore, the possibility to characterize the WM, attention, procedural memory, and information processing in terms of neurophysiological parameters will provide powerful tools for the objective and real-time evaluation of the user's mental status.

## 4.1 Working Memory

WM is considered to be one of the building blocks for higher cognitive processes (van den Bosch et al. 2014), such as decision-making (Toth and Lewis 1997), reasoning (Ruff et al. 2008), and recognition (Bledowski et al. 2012), since it is in charge of holding and simultaneously manipulating a small amount of information stored in the mind for a limited period of time for further processing (Protopapa et al. 2014). It also provides an essential interface between perception, attention, memory, and action (Baddeley 2003). WM involves three primary processes:



G. Borghini et al., *Industrial Neuroscience in Aviation*,
Biosystems & Biorobotics 18, DOI 10.1007/978-3-319-58598-7_4

encoding information, actively maintaining this information online in memory (storage), and finally, using the information to guide behavior (retrieval). During encoding, individuals actively attend and construct an internal representation of the information in memory. This mental representation of the information is maintained in a delay for a certain period, during which the information is actively prevented from decaying due to interfering or competing stimuli. Finally, the information is retrieved from the memory buffer and conveyed through a motor response (e.g., verbal, oculomotor or manual response). It is known from behavioral studies that WM performance continues to improve from childhood, through adolescence and into early adulthood (Huizinga et al. 2006; Luciana and Nelson 2002; Luna et al. 2004). The functional neurophysiological structure subserving the WM still remains controversial, but substantial progress has been made in the last years toward characterizing the neural basis of human WM (Gevins et al. 1997; Yoon et al. 2006). The anatomical substrate of WM does not appear to depend on the corticolimbic circuitry that is necessary for retrieving information when a delay of more than few seconds is interposed between study and test trials of a memory task (Cave and Squire 1992; Wickelgren 1968). Rather, lesion studies have indicated that the frontal lobes often play a critical role in this function (Frisk and Milner 1990; Owen et al. 1990). Similarly, recent neuroimaging studies have found during WM tasks metabolic increases in regions of the frontal lobes, as well as in other areas of *association cortex*, (Courtney et al. 1996; McCarthy et al. 1996). Electrophysiological methods have also been used to measure activity related to WM. Recordings of neuronal activity in primates have indicated that the neural representation of information over short delays is associated with transient activation of widespread populations of association cortex neurons (Chelazzi et al. 2001; Wilson 1993). In the humans, EEG studies have focused on *evoked potential* (EP) measures (Gevins et al. 1997; Lang et al. 1992; Ruchkin et al. 1995). For example, Gevins et al. (1997) identified phasic, subsecond changes in brain electrical signals over frontal and parietal cortex, that, in agreement with the studies of neuronal activity in primates, provided a dynamic picture of momentary changes in the spatial distribution of WM effects over the course of individual trials, as well as evidence for differences in the activity elicited between trials that have different decision requirements. Two prominent EEG features are highly sensitive to variations in WM. Several studies reported that the power of the frontal theta increased during an active maintenance of the WM information, until the information is retrieved (Jensen et al. 2002; Klimesch et al. 2005; Raghavachari et al. 2001). Fewer studies showed that the parietal and occipital theta are associated with the WM task (Kawasaki and Yamaguchi 2012; Raghavachari et al. 2006). Several studies also reported that posterior alpha suppression is associated with WM tasks (Klimesch et al. 2007; Meltzer et al. 2008). Sauseng et al. (2005) and Huang et al. (2013) reported an increase in the frontal alpha during WM tasks. The posterior regions also underwent theta increasing during a WM task, though this was followed by an alpha decrease (Lai et al. 2012).

## 4.2 Attention

The ability to suitably allocate processing resources is fundamental for the efficient processing of incoming sensory signals and for the generation of appropriate behavior in complex environments. The brain continuously receives a large amount of sensory signals that cannot be fully processed. Attention selection allows preferential processing of some signals and filtering out of other inputs. Traditionally, attention research has distinguished between two types of mechanisms that contribute to the selection process: *endogenous top-down* factors, which primarily relate to the subject's "internal" goals and intentions, and *exogenous bottom-up* factors, which primarily relate to the characteristic "external" stimulus-driven control and they are thought to take place automatically (Macaluso and Doricchi 2013). The distinction between these types of control can be rather intuitive: when searching for a friend in a crowd, we will voluntarily shift the attention from one face to another until the external input (i.e., what we see) matches our internal knowledge about the physical appearance of our friend. By contrast, in a crowd of people where everybody dressed with dark clothes, we will quickly notice a person dressed in a bright red indument, who will catch attention automatically. Many functional neuroimaging studies have investigated the neural basis of endogenous and exogenous visuospatial attention control. These ones highlighted the central role of the frontal and parietal lobes, with a notable distinction between the response patterns in dorsal and ventral regions. A variety of tasks requiring endogenous control of the spatial attention highlighted the activation of those brain regions (Corbetta et al. 1993; Gitelman et al. 1999; Yantis et al. 2002). The activation pattern typically includes the *intraparietal sulcus* (IPS) and/or the *posterior parietal cortex* (PPC) plus the *frontal eye field* (FEF) in the premotor cortex. These areas are found in serial search tasks (Fairhall et al. 2009; Himmelbach et al. 2006), and many other nonspatial attention tasks also requiring the voluntary allocation of processing resources (Kanwisher and Wojciulik 2000). Furthermore, specific experimental designs allowed segregating brain activity associated with the different phases of attention tasks, which often comprise multiple control processes (Doricchi et al. 2010; Hahn et al. 2006; Simpson et al. 2011). Sensory aspects of the incoming signals may also play an important role, particularly in the posterior parietal regions. Indeed, while showing increased activation even in the absence of any sensory input (Hopfinger et al. 2000; Kastner et al. 1999), the parietal cortex shows some residual spatially specific responses also to unattended stimuli (Saygin and Sereno 2008). The co-occurrence of sensory-driven responses and top-down internally controlled activity, also associated with motor planning, has triggered an intense debate about the format of the spatial representations in the parietal cortex (Anderson 1993; Colby and Goldberg 1999). Previous studies considered the role of the dorsal frontoparietal cortex for the processing of bottom-up signals within two main frameworks. On one hand, it has been proposed that the activation of the dorsal system "follows" the detection of relevant stimuli by the ventral attention system (Geng and Mangun 2008; Vossel et al. 2012). Accordingly to this view, the dorsal system generates top-down signals that modulate processing in the visual cortex, signals that require to reset when a relevant/infrequent "bottom-up" stimulus is presented outside the current focus of attention (Corbetta and Shulman 2002; Shulman et al. 2009). A different perspective instead entails a more "direct" activation of the dorsal system, due to forward input from the visual cortex. In this context, differences between the involvement of the posterior IPS and the anterior FEF nodes of the dorsal frontoparietal system have been proposed. For example, using concurrent *Transcranial Magnetic Stimulation* (TMS)—fMRI over either IPS or FEF, Ruff et al. (2008) showed that the interregional influence of FEF over visual cortex does not depend on the visual input, while the influence of IPS can change according to the current visual context. These findings demonstrated that in complex and naturalistic experimental situations that involve high levels of sensory competition, the dorsal frontoparietal system responds to bottom-up salient signals. In addition, despite the common assumption that mechanisms of attentional reorienting are predominantly—if not exclusively—homed in the right hemisphere (Shulman et al. 2010), a number of studies have reported a bilateral rather than just a unilateral activation of the *temporoparietal junction* (TPJ) in response to unexpected and invalid targets (Asplund et al. 2010; Bartolomeo et al. 2007; Doricchi et al. 2010; He et al. 2007). A proper level of attention is crucial in complex and operational environments: It has been demonstrated that under high workload conditions, attention appears to channel or tunnel, reducing focus on peripheral information and tasks and centralizing focus on main tasks (Staal 2004). This tunneling of attention can result in either enhanced performance or reduced performance, depending on the nature of the task and the situation. For instance, when peripheral cues are irrelevant to task completion, the ability to tune them out is likely to improve performance. On the other hand, when these peripheral cues are related to the task and their incorporation would otherwise facilitate success on the task, performance suffers when they are unattended. Furthermore, high workload might cause a reduction in environmental sampling (visual scanning) and this results in distortions in the perception of space and time (Hancock et al. 2015). Several groups have studied the effects of time pressure on attentional processes. Entin and Serfaty (1999) examined the decision-making (DM) of military personnel under the stress of time pressure in a combat simulation study. Their results suggested a reduction in the frequency and/or amount of information sought by decision-makers, as well as a reduction in the accurate detection of friend or foe submarines under high time pressure conditions. Ozel (2001) examined how people process environmental information for exiting (fire fighting scenarios) under time pressure and threat of fire. Based on the existing research literature, Ozel theorized that when slightly stressed, one's ability to determine the best time to exit was likely to be enhanced, while under high workload demand there would probably be a restriction in the range of cues expected, and a distortion of information processing. He asserted that the result was likely to be a decrement in performance. In addition, several studies support the idea that attentional tunneling can be reproduced and reversed pharmacologically. Caldwell and his colleagues (Ja and Sr 2002) have examined this line of research extensively on U.S. Army aviation personnel. For example, performance on several flight measures was enhanced after the introduction of dextroamphetamine (*Dexedrine R*): airspeed, heading, roll, and turning control. Furthermore, self-ratings of mood and physiologic measures of alertness and attention (predominantly arousal measured by EEG) also improved with pharmacological intervention. Similar findings have been obtained by the use of modafinil (*Provigil R*). This is certainly not a new assertion and many researchers have reported such findings in military aviation. For example, dextroamphetamine has been used in sustained and continuous flying operations in the military with significant success in the past (Caldwell et al. 2003). High workload-induced cognitive tunneling or narrowing of attention has also been causally linked to airline accidents and crash sequences (Staal 2004). Due to the high workload and inherent stressors associated with flying, the design of aviation systems has been a priority of various human factors engineers and cognitive scientists. Several authors have explored the relationship between HUD (*Head Up Display*) symbology used in aircraft operations and attentive processes (Dowell et al. 2002). The results of these investigations suggest that such symbology is so compelling that it fosters cognitive tunneling under the workload of flying. The result is a decrease in overall situational awareness and a greater vulnerability to error (Wickens et al. 2012). These authors have also suggested that various design modifications and compensatory strategies appeared to be available to reduce this risk. For instance, in a similar investigation, Wilson et al. (2002) found that situation-guided symbology in HUDs led to increased situational awareness, increased taxi speeds, and less workload. They surmised that this was, at least in part, due to a reduction in cognitive tunneling. Beilock et al. (2002) explored the effect of attention on sensorimotor skills. They found that well-learned skills do not require deliberate, conscious control of attention and, as a result, dual-task performance is easier. Given that well-learned skills tend to require fewer mental resources for their performance, more resources are available to devote to additional tasks accordingly. When prompted to focus attention to a particular component of the well-learned task, performance was degraded. It appears that the step-by-step attention to tasks is beneficial during initial learning stages but that this tends to be detrimental once skills are well learned.

## 4.3 Decision-Making

DM can be defined as a problem-solving activity resulting in the selection of a belief or action among alternative possibilities based on the values and preferences of the decision-maker. It is therefore a process that can be more or less rational or irrational, and can be based on explicit knowledge or tacit knowledge (Staal 2004). When facing a decision, individuals are wired to employ the most adaptive heuristic available (Gigerenzer and Selten 2002). One of the most famous heuristic is the *take-the-best* (TTB). This model proposes that decision-makers search their memories for criterion-linked probability information. Gigerenzer et al. (1991),

Bröder (2003) suggested that this search is bounded by cognitive economy. In other words, the search for information is quick, streamlined, and tends to rely on the most valid probabilistic cue separating alternatives. In other words, the authors challenged the notion that individuals reasoned from "multiple conditional probabilities," instead opting for a single discriminatory cue. When individuals are engaged in a high WM-loaded task, their capacity for other information processing operations is likely to be compromised. Bröder (2003) conjectured that, if this is so, then individuals engaged in a decision-making task would be more likely to rely on a simple heuristic than on a complex one. They would be content with limiting their scanning of alternatives, in name of a proper resource management. On the contrary, Dougherty and Hunter (2003) suggested that individuals tend to make probability judgments by comparing a focal hypothesis with relevant alternatives retrieved from long-term memory. They also concluded that individuals with a large WM span tend to include more alternatives in their comparison process, as well as that time constraints probably truncate the alternative generation process so that fewer alternatives are recalled from long-term memory, for the comparison. In general, decision-making is degraded under stressful conditions. For example, Kastner et al. (1989) employed a naval combat simulation task in their examination of the effects of task difficulty on decision-making processes. These authors designated subjects as anti-submarine warfare commanders charging them with the task of friend or foe determinations. Decision-makers were provided access (at a cost) to either a human consultant or a probe sensor. The results indicated that subjects tended to seek more information than necessary, and that such information tended to be preferentially sought from consultants who had already processed the data themselves. In an extension of this work, Entin and Serfaty (1999) explored a very similar paradigm. In their experiment, subjects were under the dual "stress" of time pressure and a secondary task that intruded on their primary decisions. The authors found that decision-makers tended to seek out more information in difficult versus easy discrimination tasks, and they typically followed a rational cost–benefit rule in their decisions. This was particularly the case, as time pressure and workload increased. Speier et al. (2003, 1999) investigated the effects of interruptions on individual decision-making, noting that interruptions make information overload worse, by reducing the amount of time that an individual has to spend working on the problem, which in turn leads to feeling time-pressured. This creates both capacity, too much information to process, and structural inputs that are occupying the same physiological channel, interference (e.g., monitoring two visual displays at once). Moreover, interruptions also place greater demand on the cognitive processing system. For example, information that is forgotten due to an overload, requires further resources to reprocess—or simply never gets encoded. Smith (1979), Damos (1991) investigated the effects of increased workload on pilots. The results indicated that periods of increased workload, even those considered to be part of normal flight operations, (1) increased the frequency and volume of errors, (2) reduced the cue sampling; (3) reduced the resource-limited capacity of WM, and (4) when time was reduced, a speed accuracy trade-off in performance outcome was adopted. Stokes (1995), Wiggins and O'Hare (1995), Wan et al. (2008) further examined pilot performance on an aeronautical decision-making task. They found that novice pilots made poorer decisions than experts, and their decision-making was further degraded under stress, while expert pilots' decision-making was not. Additionally, experienced pilots tended to visually search fewer screens of information, return to previously scanned information screens less often, and spent less time examining these screens than novice pilots. Thus, expertise in and of itself helps, but it may be that task-specific experience is most relevant to time-critical decision-making. This finding highlights the potential importance of realistic emergency and abnormal situation training, as no other way for novice operators (regardless of the industry they represent) to safely garner emergency decision-making expertise. Sperandio (1971) explored this strategy in his examination of *air traffic controllers* (ATCOs). He found that controllers tend to exercise a regulating effect on their workload through strategy shifting. Observing ATC operations, Sperandio found that controllers under increasing workload conditions tended to shift from a direct approach strategy (data needed to verify and possibly achieve separation between aircraft) to a standard approach strategy (data concerning aircraft performance not needed for separation of aircraft). Sperandio concluded that controllers economized their workload with a shift in strategy or method, reducing any redundant information or nonessential information from being processed. The neurophysiological perspective instead has demonstrated that several brain regions situated in the frontal lobes are related to the decision-making processes. Specifically, there are evidences of both the *ventromedial prefrontal cortex* (VMPFC) and the *orbitofrontal cortex* (OFC) being involved in the processing of different alternatives and potential outcomes through the assessment of their (perceived) value (Daw et al. 2006; Tremblay and Schultz 1999). In particular, the OFC is associated with the evaluation of trade-off and the expected capacity of outcomes to satisfy the user's needs (Wallis 2007). It also plays a central role in making choices about appropriate behaviors, especially when one is faced with unpredictable situations (Elliott et al. 2000). Finally, the *dorsolateral prefrontal cortex* (DLPFC) also plays a critical role in DM, as it is known for being involved in cognitive control over emotions (Rilling et al. 2008). In particular, it is involved in the control of impulses for complying with social norms, while the *ventrolateral prefrontal cortex* (VLPFC) could play a role in motivating this social norm compliance by representing the threat of punishment from others (Rilling and Sanfey 2011). Interestingly, the cognitive effort in the PFC appears to be lower when a sure gain is expected compared with risky decisions (Gonzalez et al. 2005).

## 4.4 Procedural Memory

It is widely accepted that memory consolidation processes happen off-line, after the initial hippocampus encoding event. Consolidation relies on the reactivation of neuronal circuits that are implicated in the initial encoding (Albouy et al. 2008). Experimental evidences suggest that motor memory formation occurs in two subsequent phases (Dudai 2004; Karni et al. 1994; Luft and Buitrago 2005): the first is the initial encoding of experience during training, which occurs within the first minutes-to-hours after training, and is characterized by a rapid improvement in performance. The second phase is the consolidation of memory, and involves a series of systematic changes at the molecular level (this second phase requires a longer time). During the consolidation phase, memories are reorganized, and the hippocampus-dependent initial memories may become hippocampal-independent (Albouy et al. 2008). Processes of reactivation of memories lead to renewed consolidation each time that the reactivations occur, thus enhancing the first consolidated memory representation and converting it into a long-lasting stable memory trace (Dudai 2004). Delayed additional gains occur after the second phase, even without additional practice (Karni et al. 1994). Several lines of evidence suggest that the theta oscillations play an important role in the formation of memory: theta oscillations are typical of the hippocampal activity, upon memory encoding. The latter generates oscillations that can propagate to other brain structures (even relatively distant), which support memory consolidation and are thought to play a critical role in the induction of long-term plasticity, in association with the consolidation of memory (Chauvette 2013; Kropotov 2010). Theta rhythms are correlated with the episodic and semantic memory (Buzsáki 2005; Guderian and Düzel 2005; Raghavachari et al. 2006) and are involved in learning and memory within the mPFC (medial prefrontal cortex), as well as the hippocampal system (Anderson et al. 2010; Benchenane et al. 2010). Several studies point to the theta synchronization as a mechanism that, during the consolidation, underlies the communication between the hippocampus, the VMPFC and some remote memory areas. The underlying mechanism is still not clear, but on the other side, there has been an attempt to explain the mechanism of memory consolidation, which is known as the "system-level memory consolidation theory" (Nieuwenhuis and Takashima 2011). This model suggests that, whereas the hippocampus is strongly activated in the first stages of memory related to the neocortical formations, new forms of memory gradually become independent from the hippocampal activations, and consolidation correlates with an increased activation in the human subgenual VMPFC. The vmPFC, similar to the anterior cingulate cortex, seems to link the neocortical representational areas in the remote memory (Nieuwenhuis and Takashima 2011). The system-level memory consolidation view implies an exchange of information in a network of brain areas centered on the hippocampus and articulated with communicating areas, which include the neocortex and structures like the amygdala and the striatum (Battaglia et al. 2011). The interaction between the hippocampus and the striatum resembles the interaction between the hippocampus and neocortex (Battaglia et al. 2011). This exchange is theorized to be linked to the theta oscillations: hippocampal cells preferentially fire at a specific theta phase (Klausberger et al. 2003; Mizuseki et al. 2009), and so do the areas in the medial temporal, the parietal lobes, and the rest of the areas that exchange information with the hippocampus. Thus, theta is assumed to regulate information exchange between the hippocampus and the striatum (for a review see Battaglia et al. 2011). This exchange of information extends to relatively distant sensory and associative areas of the parietal cortex, which are also entrained by theta oscillations (Sirota et al. 2008). This exchange of information is based on a dynamical evolving schema where the theta oscillations orchestrate a synchronized discharge of cell assemblations across brain structures in order to encode information. Recent results further support the central role of the hippocampus–striatum exchange in motor memory consolidation, and suggest that the interplay between the striatum and the hippocampus, during motor training conditions subsequent motor sequence memory consolidation, which is further supported by reorganization of cerebral activity in hippocampo-neocortical networks after sleep (Albouy et al. 2013). The ventral striatum is involved in learning beyond memory consolidation, and was found to be related to individual variations in learning performance (Vink et al. 2013). We hypothesize that enhanced theta supports exchange of information between the hippocampus and the neocortical areas during the consolidation of memory, hence it will be reflected in indicators of memory consolidation (Gruzelier 2009; Gruzelier et al. 2006). Gruzieler (2009) associated increased parietal theta (recorded from the Pz channel) with increased activity in the hippocampus. Parietal theta synchronization was also found to be correlated with retrieval (Jacobs et al. 2006; Sauseng et al. 2005). In addition, the system-level theory predicts that synchronized elevated theta power leads to consolidation of memory. To test the role of theta in memory consolidation, Reiner et al. (2014) tried to understand whether enhanced theta, during awake-hours, affects or not the consolidation of memory. Indeed, results show that enhanced theta is correlated with behavioral changes indicating consolidation of memory.

## References

Albouy G, Sterpenich V, Balteau E, Vandewalle G, Desseilles M, Dang-Vu T, Darsaud A, Ruby P, Luppi P-H, Degueldre C, Peigneux P, Luxen A, Maquet P (2008) Both the hippocampus and striatum are involved in consolidation of motor sequence memory. Neuron 58:261–272. doi:10.1016/j.neuron.2008.02.008

Albouy G, King BR, Maquet P, Doyon J (2013) Hippocampus and striatum: dynamics and interaction during acquisition and sleep-related motor sequence memory consolidation. Hippocampus 23:985–1004. doi:10.1002/hipo.22183

Anderson JR (1993) Rules of the mind. Psychology Press

Anderson KL, Rajagovindan R, Ghacibeh GA, Meador KJ, Ding M (2010) Theta oscillations mediate interaction between prefrontal cortex and medial temporal lobe in human memory. Cereb Cortex 1991(20):1604–1612. doi:10.1093/cercor/bhp223

Asplund CL, Todd JJ, Snyder AP, Marois R (2010) A central role for the lateral prefrontal cortex in goal-directed and stimulus-driven attention. Nat Neurosci 13:507–512. doi:10.1038/nn.2509

Baddeley A (2003) Working memory: looking back and looking forward. Nat Rev Neurosci 4:829–839. doi:10.1038/nrn1201

Bartolomeo P, Thiebaut de Schotten M, Doricchi F (2007) Left unilateral neglect as a disconnection syndrome. Cereb Cortex 17:2479–2490. doi:10.1093/cercor/bhl181

Battaglia FP, Benchenane K, Sirota A, Pennartz CMA, Wiener SI (2011) The hippocampus: hub of brain network communication for memory. Trends Cogn Sci 15:310–318. doi:10.1016/j.tics.2011.05.008

Beilock SL, Carr TH, MacMahon C, Starkes JL (2002) When paying attention becomes counterproductive: impact of divided versus skill-focused attention on novice and experienced performance of sensorimotor skills. J Exp Psychol Appl 8:6–16. doi:10.1037/1076-898X.8.1.6
Benchenane K, Peyrache A, Khamassi M, Tierney PL, Gioanni Y, Battaglia FP, Wiener SI (2010) Coherent theta oscillations and reorganization of spike timing in the hippocampal-prefrontal network upon learning. Neuron 66:921–936. doi:10.1016/j.neuron.2010.05.013
Bledowski C, Kaiser J, Wibral M, Yildiz-Erzberger K, Rahm B (2012) Separable neural bases for subprocesses of recognition in working memory. Cereb Cortex 22:1950–1958. doi:10.1093/cercor/bhr276
Bröder A (2003) Decision making with the "adaptive toolbox": Influence of environmental structure, intelligence, and working memory load. J Exp Psychol Learn Mem Cogn 29: 611–625. doi:10.1037/0278-7393.29.4.611
Buzsáki G (2005) Theta rhythm of navigation: link between path integration and landmark navigation, episodic and semantic memory. Hippocampus 15:827–840. doi:10.1002/hipo.20113
Caldwell JA, Caldwell JL, Darlington KK (2003) Utility of dextroamphetamine for attenuating the impact of sleep deprivation in pilots. Aviat Space Environ Med 74:1125–1134
Cave CB, Squire LR (1992) Intact verbal and nonverbal short-term memory following damage to the human hippocampus. Hippocampus 2:151–163. doi:10.1002/hipo.450020207
Chauvette S (2013) Slow-wave sleep: generation and propagation of slow waves, role in long-term plasticity and gating. Université Laval
Chelazzi L, Miller EK, Duncan J, Desimone R (2001) Responses of neurons in macaque area V4 during memory-guided visual search. Cereb Cortex 11:761–772. doi:10.1093/cercor/11.8.761
Colby CL, Goldberg ME (1999) Space and attention in parietal cortex. Annu Rev Neurosci 22:319–349. doi:10.1146/annurev.neuro.22.1.319
Corbetta M, Shulman GL (2002) Control of goal-directed and stimulus-driven attention in the brain. Nat Rev Neurosci 3:201–215. doi:10.1038/nrn755
Corbetta M, Miezin FM, Shulman GL, Petersen SE (1993) A PET study of visuospatial attention. J Neurosci 13:1202–1226
Courtney SM, Ungerleider LG, Keil K, Haxby JV (1996) Object and spatial visual working memory activate separate neural systems in human cortex. Cereb Cortex 6:39–49. doi:10.1093/cercor/6.1.39
Damos D (1991) Multiple task performance. CRC Press, Florida
Daw ND, O'Doherty JP, Dayan P, Seymour B, Dolan RJ (2006) Cortical substrates for exploratory decisions in humans. Nature 441:876–879. doi:10.1038/nature04766
Doricchi F, Macci E, Silvetti M, Macaluso E (2010) Neural correlates of the spatial and expectancy components of endogenous and stimulus-driven orienting of attention in the posner task. Cereb Cortex 20:1574–1585. doi:10.1093/cercor/bhp215
Dougherty MRP, Hunter J (2003) Probability judgment and subadditivity: the role of working memory capacity and constraining retrieval. Mem Cognit 31:968–982. doi:10.3758/BF03196449
Dowell SR, Foyle DC, Hooey BL, Williams JL (2002) The effect of visual location on cognitive tunneling with superimposed hud symbology. In: Proceedings of the human factors and ergonomics society annual meeting, vol 46. pp 121–125. doi:10.1177/154193120204600125
Dudai Y (2004) The neurobiology of consolidations, or, how stable is the engram? Annu Rev Psychol 55:51–86. doi:10.1146/annurev.psych.55.090902.142050
Elliott R, Dolan RJ, Frith CD (2000) Dissociable functions in the medial and lateral orbitofrontal cortex: evidence from human neuroimaging studies. Cereb Cortex 10:308–317. doi:10.1093/cercor/10.3.308
Entin EE, Serfaty D (1999) Adaptive team coordination. Hum Factors 41:312–325. doi:10.1518/001872099779591196
Fairhall SL, Indovina I, Driver J, Macaluso E (2009) The brain network underlying serial visual search: comparing overt and covert spatial orienting, for activations and for effective connectivity. Cereb Cortex 19:2946–2958. doi:10.1093/cercor/bhp064

Foyle DC, Dowell SR, Hooey BL (n.d.) Cognitive tunneling in head-up display (hud) superimposed symbology: effects of information location
Frisk V, Milner B (1990) The role of the left hippocampal region in the acquisition and retention of story content. Neuropsychologia 28:349–359
Geng JJ, Mangun GR (2008) Anterior intraparietal sulcus is sensitive to bottom–up attention driven by stimulus salience. J Cogn Neurosci 21:1584–1601. doi:10.1162/jocn.2009.21103
Gevins A, Smith ME, McEvoy L, Yu D (1997) High-resolution EEG mapping of cortical activation related to working memory: effects of task difficulty, type of processing, and practice. Cereb Cortex 1991(7):374–385
Gigerenzer G, Selten R (2002) Bounded rationality: the adaptive toolbox. MIT Press, Cambridge
Gigerenzer G, Hoffrage U, Kleinbölting H (1991) Probabilistic mental models: a Brunswikian theory of confidence. Psychol Rev 98:506–528
Gitelman DR, Nobre AC, Parrish TB, LaBar KS, Kim Y-H, Meyer JR, Mesulam M-M (1999) A large-scale distributed network for covert spatial attention. Further anatomical delineation based on stringent behavioural and cognitive controls. Brain 122:1093–1106. doi:10.1093/brain/122.6.1093
Gonzalez C, Dana J, Koshino H, Just M (2005) The framing effect and risky decisions: examining cognitive functions with fMRI. J Econ Psychol 26:1–20. doi:10.1016/j.joep.2004.08.004
Gruzelier J (2009) A theory of alpha/theta neurofeedback, creative performance enhancement, long distance functional connectivity and psychological integration. Cogn Process 10(Suppl 1): S101–S109. doi:10.1007/s10339-008-0248-5
Gruzelier J, Egner T, Vernon D (2006) Validating the efficacy of neurofeedback for optimising performance. Prog Brain Res 159:421–431. doi:10.1016/S0079-6123(06)59027-2
Guderian S, Düzel E (2005) Induced theta oscillations mediate large-scale synchrony with mediotemporal areas during recollection in humans. Hippocampus 15:901–912. doi:10.1002/hipo.20125
Hahn B, Ross TJ, Stein EA (2006) Neuroanatomical dissociation between bottom–up and top–down processes of visuospatial selective attention. NeuroImage 32:842–853. doi:10.1016/j.neuroimage.2006.04.177
Hancock PA, Sawyer BD, Stafford S (2015) The effects of display size on performance. Ergonomics 58:337–354. doi:10.1080/00140139.2014.973914
Harris D (ed) (2015) Engineering psychology and cognitive ergonomics., Lecture notes in computer scienceSpringer International Publishing, Cham
He BJ, Snyder AZ, Vincent JL, Epstein A, Shulman GL, Corbetta M (2007) Breakdown of functional connectivity in frontoparietal networks underlies behavioral deficits in spatial neglect. Neuron 53:905–918. doi:10.1016/j.neuron.2007.02.013
Himmelbach M, Erb M, Karnath H-O (2006) Exploring the visual world: the neural substrate of spatial orienting. NeuroImage 32:1747–1759. doi:10.1016/j.neuroimage.2006.04.221
Hopfinger JB, Buonocore MH, Mangun GR (2000) The neural mechanisms of top-down attentional control. Nat Neurosci 3:284–291. doi:10.1038/72999
Huang L-Y, She H-C, Chou W-C, Chuang M-H, Duann J-R, Jung T-P (2013) Brain oscillation and connectivity during a chemistry visual working memory task. Int J Psychophysiol 90: 172–179. doi:10.1016/j.ijpsycho.2013.07.001
Huizinga M, Dolan CV, van der Molen MW (2006) Age-related change in executive function: developmental trends and a latent variable analysis. Neuropsychologia 44:2017–2036. doi:10.1016/j.neuropsychologia.2006.01.010
Ja C, Sr G (2002) A survey of aircrew fatigue in a sample of U.S. Army aviation personnel. Aviat Space Environ Med 73:472–480
Jacobs J, Hwang G, Curran T, Kahana MJ (2006) EEG oscillations and recognition memory: theta correlates of memory retrieval and decision making. NeuroImage 32:978–987. doi:10.1016/j.neuroimage.2006.02.018
Jensen O, Gelfand J, Kounios J, Lisman JE (2002) Oscillations in the alpha band (9–12 Hz) increase with memory load during retention in a short-term memory task. Cereb Cortex 1991 (12):877–882

Kanwisher N, Wojciulik E (2000) Visual attention: insights from brain imaging. Nat Rev Neurosci 1:91–100. doi:10.1038/35039043

Karni A, Tanne D, Rubenstein BS, Askenasy JJ, Sagi D (1994) Dependence on REM sleep of overnight improvement of a perceptual skill. Science 265:679–682

Kastner MP, Entin EE, Castanon DA, Serfaty D, Deckert JC (1989) A normative-descriptive study of team detection with communication alternatives. In: Presented at the conference proceedings, IEEE international conference on systems, man and cybernetics, vol 2. pp 699–705. doi:10.1109/ICSMC.1989.71385

Kastner S, Pinsk MA, De Weerd P, Desimone R, Ungerleider LG (1999) Increased activity in human visual cortex during directed attention in the absence of visual stimulation. Neuron 22:751–761. doi:10.1016/S0896-6273(00)80734-5

Kawasaki M, Yamaguchi Y (2012) Individual visual working memory capacities and related brain oscillatory activities are modulated by color preferences. Front Hum Neurosci 6. doi:10.3389/fnhum.2012.00318

Klausberger T, Magill PJ, Márton LF, Roberts JDB, Cobden PM, Buzsáki G, Somogyi P (2003) Brain-state- and cell-type-specific firing of hippocampal interneurons in vivo. Nature 421: 844–848. doi:10.1038/nature01374

Klimesch W, Schack B, Sauseng P (2005) The functional significance of theta and upper alpha oscillations. Exp Psychol 52:99–108

Klimesch W, Sauseng P, Hanslmayr S (2007) EEG alpha oscillations: the inhibition-timing hypothesis. Brain Res Rev 53:63–88. doi:10.1016/j.brainresrev.2006.06.003

Kropotov JD (2010) Quantitative EEG, event-related potentials and neurotherapy. Academic Press, Cambridge

Lai K, She H-C, Chen S-C, Chou W-C, Huang L-Y, Jung T-P, Gramann K (2012) Encoding of physics concepts: concreteness and presentation modality reflected by human brain dynamics. PLoS ONE 7:e41784. doi:10.1371/journal.pone.0041784

Lang W, Starr A, Lang V, Lindinger G, Deecke L (1992) Cortical DC potential shifts accompanying auditory and visual short-term memory. Electroencephalogr Clin Neurophysiol 82:285–295

Luciana M, Nelson CA (2002) Assessment of neuropsychological function through use of the Cambridge neuropsychological testing automated battery: performance in 4-to 12-year-old children. Dev Neuropsychol 22:595–624. doi:10.1207/S15326942DN2203_3

Luft AR, Buitrago MM (2005) Stages of motor skill learning. Mol Neurobiol 32:205–216. doi:10.1385/MN:32:3:205

Luna B, Garver KE, Urban TA, Lazar NA, Sweeney JA (2004) Maturation of cognitive processes from late childhood to adulthood. Child Dev 75:1357–1372. doi:10.1111/j.1467-8624.2004.00745.x

Macaluso E, Doricchi F (2013) Attention and predictions: control of spatial attention beyond the endogenous-exogenous dichotomy. Front Hum Neurosci 7:685. doi:10.3389/fnhum.2013.00685

McCarthy G, Puce A, Constable T, Krystal JH, Gore JC, Goldman-Rakic P (1996) Activation of human prefrontal cortex during spatial and nonspatial working memory tasks measured by functional MRI. Cereb Cortex 6:600–611. doi:10.1093/cercor/6.4.600

Meltzer JA, Zaveri HP, Goncharova II, Distasio MM, Papademetris X, Spencer SS, Spencer DD, Constable RT (2008) Effects of working memory load on oscillatory power in human intracranial EEG. Cereb Cortex 18:1843–1855. doi:10.1093/cercor/bhm213

Mizuseki K, Sirota A, Pastalkova E, Buzsáki G (2009) Theta oscillations provide temporal windows for local circuit computation in the entorhinal-hippocampal loop. Neuron 64: 267–280. doi:10.1016/j.neuron.2009.08.037

Nieuwenhuis ILC, Takashima A (2011) The role of the ventromedial prefrontal cortex in memory consolidation. Behav Brain Res 218:325–334. doi:10.1016/j.bbr.2010.12.009

Owen AM, Downes JJ, Sahakian BJ, Polkey CE, Robbins TW (1990) Planning and spatial working memory following frontal lobe lesions in man. Neuropsychologia 28:1021–1034

Ozel F (2001) Time pressure and stress as a factor during emergency egress. Saf Sci 38:95–107. doi:10.1016/S0925-7535(00)00061-8

Protopapa F, Siettos CI, Evdokimidis I, Smyrnis N (2014) Granger causality analysis reveals distinct spatio-temporal connectivity patterns in motor and perceptual visuo-spatial working memory. Front Comput Neurosci 8. doi:10.3389/fncom.2014.00146

Raghavachari S, Kahana MJ, Rizzuto DS, Caplan JB, Kirschen MP, Bourgeois B, Madsen JR, Lisman JE (2001) Gating of human theta oscillations by a working memory task. J Neurosci Off J Soc Neurosci 21:3175–3183

Raghavachari S, Lisman JE, Tully M, Madsen JR, Bromfield EB, Kahana MJ (2006) Theta oscillations in human cortex during a working-memory task: evidence for local generators. J Neurophysiol 95:1630–1638. doi:10.1152/jn.00409.2005

Reiner M, Rozengurt R, Barnea A (2014) Better than sleep: theta neurofeedback training accelerates memory consolidation. Biol Psychol 95:45–53. doi:10.1016/j.biopsycho.2013.10.010

Rilling JK, Sanfey AG (2011) The neuroscience of social decision-making. Annu Rev Psychol 62:23–48. doi:10.1146/annurev.psych.121208.131647

Rilling JK, Dagenais JE, Goldsmith DR, Glenn AL, Pagnoni G (2008) Social cognitive neural networks during in-group and out-group interactions. NeuroImage 41:1447–1461. doi:10.1016/j.neuroimage.2008.03.044

Ruchkin DS, Canoune HL, Johnson R, Ritter W (1995) Working memory and preparation elicit different patterns of slow wave event-related brain potentials. Psychophysiology 32:399–410. doi:10.1111/j.1469-8986.1995.tb01223.x

Ruff CC, Bestmann S, Blankenburg F, Bjoertomt O, Josephs O, Weiskopf N, Deichmann R, Driver J (2008) Distinct causal influences of parietal versus frontal areas on human visual cortex: evidence from concurrent TMS–fMRI. Cereb Cortex 18:817–827. doi:10.1093/cercor/bhm128

Sauseng P, Klimesch W, Schabus M, Doppelmayr M (2005) Fronto-parietal EEG coherence in theta and upper alpha reflect central executive functions of working memory. Int J Psychophysiol Off J Int Organ Psychophysiol 57:97–103. doi:10.1016/j.ijpsycho.2005.03.018

Saygin AP, Sereno MI (2008) Retinotopy and attention in human occipital, temporal, parietal, and frontal cortex. Cereb Cortex 18:2158–2168. doi:10.1093/cercor/bhm242

Shulman GL, Astafiev SV, Franke D, Pope DLW, Snyder AZ, McAvoy MP, Corbetta M (2009) Interaction of stimulus-driven reorienting and expectation in ventral and dorsal frontoparietal and basal ganglia-cortical networks. J Neurosci 29:4392–4407. doi:10.1523/JNEUROSCI.5609-08.2009

Shulman GL, Pope DLW, Astafiev SV, McAvoy MP, Snyder AZ, Corbetta M (2010) Right hemisphere dominance during spatial selective attention and target detection occurs outside the dorsal frontoparietal network. J Neurosci 30:3640–3651. doi:10.1523/JNEUROSCI.4085-09.2010

Simpson GV, Weber DL, Dale CL, Pantazis D, Bressler SL, Leahy RM, Luks TL (2011) Dynamic activation of frontal, parietal, and sensory regions underlying anticipatory visual spatial attention. J Neurosci 31:13880–13889. doi:10.1523/JNEUROSCI.1519-10.2011

Sirota A, Montgomery S, Fujisawa S, Isomura Y, Zugaro M, Buzsáki G (2008) Entrainment of neocortical neurons and gamma oscillations by the hippocampal theta rhythm. Neuron 60:683–697. doi:10.1016/j.neuron.2008.09.014

Smith HPR (1979) A simulator study of the interaction of pilot workload with errors, vigilance, and decisions

Speier C, Valacich JS, Vessey I (1999) The influence of task interruption on individual decision making: an information overload perspective. Decis Sci 30:337–360. doi:10.1111/j.1540-5915.1999.tb01613.x

Speier C, Vessey I, Valacich JS (2003) The effects of interruptions, task complexity, and information presentation on computer-supported decision-making performance. Decis Sci 34:771–797. doi:10.1111/j.1540-5414.2003.02292.x

Sperandio JC (1971) Variation of operator's strategies and regulating effects on workload. Ergonomics 14:571–577. doi:10.1080/00140137108931277

Staal MA (2004) Stress, cognition, and human performance: a literature review and conceptual framework
Stokes AF (1995) Sources of stress-resistant performance in aeronautical decision making: the role of knowledge representation and trait anxiety. Proc Hum Factors Ergon Soc Annu Meet 39:887–890. doi:10.1177/154193129503901307
THE DISTORTION OF PERCEPTUAL SPACE-TIME UNDER STRESS (PDF Download Available) [WWW Document], (n.d.) ResearchGate. https://www.researchgate.net/publication/265893566_THE_DISTORTION_OF_PERCEPTUAL_SPACE-TIME_UNDER_STRESS. Accessed 3 Apr 2017
Toth JA, Lewis CM (1997) The role of working memory and external representation in individual decision making (AAAI Technical Report No. FS-97-03)
Tremblay L, Schultz W (1999) Relative reward preference in primate orbitofrontal cortex. Nature 398:704–708. doi:10.1038/19525
van den Bosch GE, Marroun HE, Schmidt MN, Tibboel D, Manoach DS, Calhoun VD, White TJH (2014) Brain connectivity during verbal working memory in children and adolescents. Hum Brain Mapp 35:698–711. doi:10.1002/hbm.22193
Vink M, Pas P, Bijleveld E, Custers R, Gladwin TE (2013) Ventral striatum is related to within-subject learning performance. Neuroscience 250:408–416. doi:10.1016/j.neuroscience.2013.07.034
Vossel S, Weidner R, Driver J, Friston KJ, Fink GR (2012) Deconstructing the architecture of dorsal and ventral attention systems with dynamic causal modeling. J Neurosci 32:10637–10648. doi:10.1523/JNEUROSCI.0414-12.2012
Wallis JD (2007) Orbitofrontal cortex and its contribution to decision-making. Annu Rev Neurosci 30:31–56. doi:10.1146/annurev.neuro.30.051606.094334
Wan L, Van Diest I, De Peuter S, Bogaerts K, Oyen N, Hombroux N, Van de Woestijne K, Gallego J, Van den Bergh O (2008) Repeated experiences of air hunger and ventilatory behavior in response to hypercapnia in the standardized rebreathing test: effects of anxiety. Biol Psychol 77:223–232. doi:10.1016/j.biopsycho.2007.10.013
Wickelgren WA (1968) Sparing of short-term memory in an amnesic patient: implications for strength theory of memory. Neuropsychologia 6:235–244. doi:10.1016/0028-3932(68)90022-5
Wickens CD, Hollands JG, Banbury S, Parasuraman R (2012) Engineering psychology & human performance, 4th edn. Psychology Press, Boston
Wiggins M, O'Hare D (1995) Expertise in aeronautical weather-related decision making: a cross-sectional analysis of general aviation pilots. J Exp Psychol Appl 1:305–320. doi:10.1037/1076-898X.1.4.305
Wilson GF (1993) Air-to-ground training missions: a psychophysiological workload analysis. Ergonomics 36:1071–1087. doi:10.1080/00140139308967979
Wilson GF (2002) An analysis of mental workload in pilots during flight using multiple psycho-physiological measures. Int J Aviat Psychol 3–18
Yantis S, Schwarzbach J, Serences JT, Carlson RL, Steinmetz MA, Pekar JJ, Courtney SM (2002) Transient neural activity in human parietal cortex during spatial attention shifts. Nat Neurosci 5:995–1002. doi:10.1038/nn921
Yoon JH, Curtis CE, D'Esposito M (2006) Differential effects of distraction during working memory on delay-period activity in the prefrontal cortex and the visual association cortex. NeuroImage 29:1117–1126. doi:10.1016/j.neuroimage.2005.08.024

# Chapter 5
# Simulators

## 5.1 Simulators

This chapter aims to describe the contexts and technology by which the experimental scenarios have been performed and designed. The general approach of the studies presented in the book started from very controlled settings to realistic environments. From scientific literature reviews concerning the phenomena under investigation, specific and controlled tasks were selected and used to test the developed methodologies. The final goal was to define and tune neurometrics able to provide objective information regarding changes of the considered cognitive phenomena along the execution of the tasks. Controlled tasks are usually simple and easy to modify in order to elicit specific reactions and behaviors of the participants. Generally, they require to sit in front of a monitor and answers the stimulus by pressing button or moving a joystick. On the contrary, it is not always possible to keep under control realistic activities as piloting a jet fighter or managing several aircrafts. Together with subject matter experts (SMEs), we designed missions able to simulate, with high fidelity, ecological conditions in which several phases and variables were under control. This capability allowed us to perform and test our methodologies in realistic contexts by recruiting professional personnel. In the following sections, we illustrate some of the laboratory tasks and ecological settings used in our works.

## 5.2 NASA—Multi Attribute Task Battery (MATB)

The NASA—*Multi Attribute Task Battery* (MATB) is a computer-based platform designed by the NASA to evaluate pilots' and operators' performance and workload (Comstock 1994). It could be freely downloaded from the NASA Website at the following URL: http://matb.larc.nasa.gov/. It is a standard platform for the evaluation of the cognitive operational capability, since it is able to provide different



G. Borghini et al., *Industrial Neuroscience in Aviation*,
Biosystems & Biorobotics 18, DOI 10.1007/978-3-319-58598-7_5

tasks that have to be attended by the participant in parallel, as well as single tasks that can be widely modulated in difficulty (Fig. 5.1).

In particular, the experimental software requires the simultaneous performance of tasks that are generalizations of piloting tasks: tracking (TRCK), auditory monitoring (COMM), resource management (RMAN), and response to event onsets (SYSM). The demand of the system monitoring task is to monitor the gauges and the warning lights by responding to the absence of the Green light and/or the presence of the Red light, and to monitor the four moving pointer dials for deviation from the midpoint. The demands of manual control are simulated by the tracking task: the participant keeps the target at the center of the window by moving the joystick, and his task can be automated to simulate the reduced manual demands of the autopilot. Participants are also required to respond to simulated communication tasks, which consist in prerecorded auditory messages to the operator at selected intervals during the simulation. However, not all the messages are relevant to the operator, so the goal of the COMM task is to determine which messages are relevant, and then to respond by selecting the appropriate radio and frequency on the communication task's window. The demands of fuel management are simulated by the resource management task: the goal is to maintain the main tanks at 2500 (lbs). This is done by turning *On* or *Off* any of the eight pumps, all of which become red when they fail. At the end of the task execution, four performance indexes

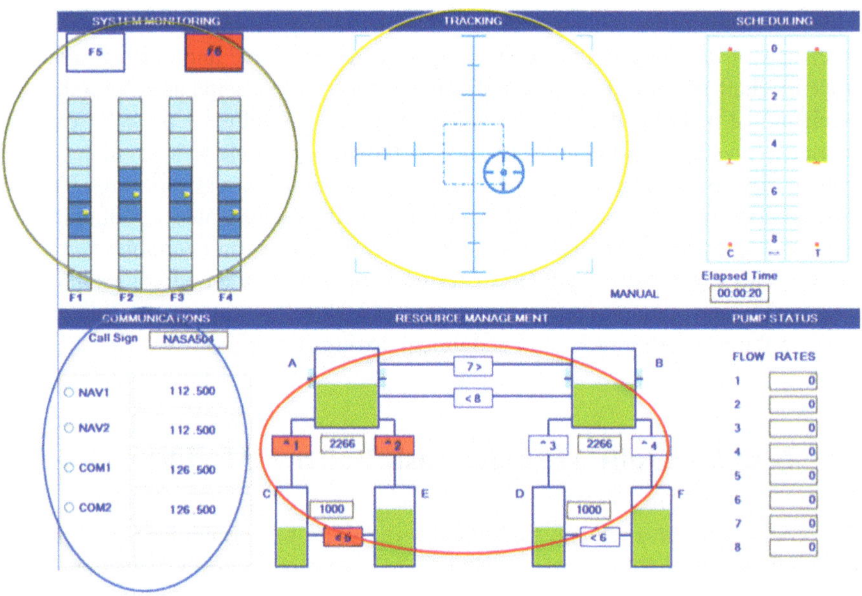


**Fig. 5.1** NASA's —Multi Attribute Task Battery (MATB) interface. On the top left corner (*gree circle*), there is the emergency lights subtask; on the top, in the center (*yellow circle*), there is the cursor tracking task; on the left bottom corner (*blue circle*), there is the radio communication task and, finally, in the center on the bottom (*red circle*), there is the fuel managing task

(one for each subtask) have been defined by means of the log files provided by the MATB. The index for the TRCK task has been defined by considering the complement of the ratio between the cursor's distance got by the participant and the maximum of this distance (fixed) from the center of the screen. The indexes of the COMM and SYSM tasks have been defined as a linear combination of accuracies in terms of correct answers (e.g., correct radio or frequency selected) and the complement of the ratio between the participant's reaction time and the maximum time for answering; then, the results have been multiplied for "100" in order to obtain a percentage. Finally, the index for the RMAN task has been defined as the mean value of the fuel's levels in the tank A and B, multiplied by "100". In order to get a global performance index, the average of the previous indexes has been calculated, as the single index was a percentage.

## 5.3 ATM Environment

*Air Traffic Management* (ATM) is a term largely used to encompass all the systems that assist aircrafts departing from an aerodrome, the transit air-space, and their land at destination. It includes different areas such as *Air Traffic Control* (ATC), *Air Traffic Flow Management* (ATFM), and *Aeronautical Information Services* (AIS). Rather than discussing all these aspects of the Air Traffic Management Systems, the following studies will focus on the area of ATC, in which an *Air Traffic Controller* (ATCO) monitors the movement of a number of aircrafts in his area of responsibility—typically via flight progress strips and a radar screen—and provides control instructions to the pilots (if needed, e.g., in case of risk of violation of the legal safety separations between aircraft). Compared to other safety critical and high-hazard domains, such as nuclear power, air traffic control is characterized by the key role played by human actors. Despite advances in technology, the ATM system is still human-centerd. As a matter of fact, safety relevant decisions are taken mostly by humans, whereas computer systems are supporting tools whose purpose is to assist the controller in monitoring and communication tasks. In the review about the levels of automation, Parasuraman et al. (2000) argued that safety critical systems, such as air traffic control system, are typically less automated. Automation might be higher in information acquisition and information analysis tasks (provided that the automation is reliable) and lower for decision selection, whereas moderately high for action implementation. Indeed, in ATM systems, the controller's work is very demanding—in a cognitive sense. Activities for managing traffic—such as solving conflicts, maintaining separations between aircrafts and coordinating traffic —involve cognitive processes such as visual scanning, managing situation awareness, decision-making and attention control. Controllers have a key role in facing system complexity, because their main objective is to anticipate and manage unpredictable situations affected by multiple elements. Complexity, in fact, does not regard only the environment: radio and phone communications, phone communications, radar displays and computers are all system elements adding complexity.

A peculiar characteristic of the ATM work is that ATCOs usually perform few recurring tasks—apart maybe from routine tasks (i.e., welcoming aircraft on the frequency and handing them off to adjacent sectors). Anyway, even if these are well-known tasks, their order remains largely unpredictable because of the dynamic nature of weather, traffic demands, operational conditions, and so on. ATC research could benefit greatly from psyco-physiological approaches, i.e., the ability to measure certain physiological indicators known to be linked to underlying mental and physiological processes (since the work of air traffic controllers involves little physical activity, mental processes will be the main concern, but physiological artifacts are of some concern, e.g., eyeblink artifacts on certain EEG signals). Among these psychophysiological approaches, neurometrics like measuring the brains' electrical activity with the aim to infer the brain activity itself, are certainly very promising. In this regard, specific experiments and *ad hoc* ATM scenarios have been designed by expert Controllers from the Training Centre of the *École Nationale de l'Aviation Civile* (ENAC) of Toulouse (France). In addition, professional and students *en-route* Controllers have been recruited and involved in the experiments byENAC. The creation of the ATM tasks has started from a scenario already used for training, and it has been modified according to the experiments' objectives and needs. The adjustments mostly impacted the routes and air traffic within the air-space sector (adding or changing them), while the borders of the sectors have not been changed. The air traffic adjustments has consisted in flights rescheduling and flight levels' (FLs) changing. Three ATM scenarios have been defined according to the *Number of aircraft*, *Traffic geometry*, and *Number of conflicts*.

The scenarios have been created with different air traffic geometries but similar difficulty levels, and each of them has been defined as a combination of three air traffic level: *Easy* (E), *Medium* (M), and *Hard* (H). The total duration of the scenario was 45 min, while the easy, medium and hard levels lasted 15 min. During the execution of the ATM task, the controller may assume an aircraft before it enters the sector and could hand over an aircraft before it actually leaves the sector. In other words, the ATCO could anticipate the heavy traffic loads when they are still in the easy period and, consequently, change the duration of the next condition.

The considered air-space has been featured by an *en-route* sector, where the air traffic has been handled and transferred according to the standard protocols currently used in the Area Control Center (ACC) of Toulouse. Figure 5.2 shows the sector features. The RU (dark gray) is the considered air-sector with routes and beacons (light gray). The radio frequency is FM 125,305 and the controlled altitude range starts from FL335 to above. The RS sector is under RU sector, and air traffic between those sectors has to respect specific procedures in terms of altitude levels, when flying through the RS sector—depending on departure or arrival flights. Specifically, departing flights through the RS sector have to climb to FL330 before crossing RU, while arrival flights through the RS sectors have to descent to FL340 before entering the RU air-space. Also, *feeder sectors* have been simulated with the aim to handover traffics within the considered sectors, and they are labeled in Fig. 5.2 as *OS*, *OM*, *N3*, *G2*, *M2*, *I3*, *I2*, *S3,* and *S2*. The red and green arrows show the traffic flow direction, respectively, incoming and outgoing traffic.

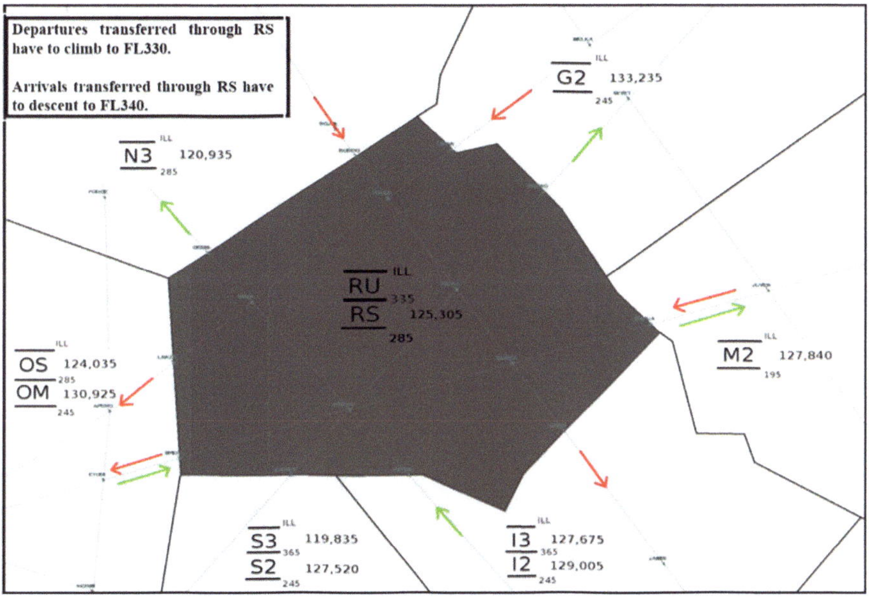


**Fig. 5.2** The RU (*dark gray*) is the considered air-sector with routes and beacons (*light gray*). The controlled altitude range starts from FL335 to above. The RS sector is under RU sector and air traffic between those sectors is described in the box in the top left corner. The feeder sectors are OS, OM, N3, G2, M2, I3, I2, S3 and S2. The *red* and *green arrows* show the traffic flow direction, respectively, incoming and outcoming traffic

The operational ATC position (left side in Fig. 5.3) consisted in two screens: one screen of 30″ (Radar Image), and a second one of 21″ (tactile WACOM interface to interact with the radar image (zoom, move, clearances, and information). The simulated ATC position was very similar to the French *Control Working Position* (CWP) in real ATM environments, except for the interface. The WACOM interface was a prototype where the ATCO could interact with the RADAR by means of a touch-screen and capacitive-pen. Before the experiments, the ATCOs have been trained to use it correctly in order to avoid learning effects during the recordings of brain activity.

To make the ATC simulation realistic, two Pseudo Pilots have been involved in the experiments with the aim to have real communications with the ATCOs. Headsets (headphone and microphone) have been used by the ATCOs and Pseudo Pilots. The interface of the pilots is reported in Fig. 5.4 and provided information about all the airplanes on the RADAR and those assumed by the ATCOs (in particular, the main view shows the selected aircraft navigation display). On the left side, the managed aircraft are listed, while the second list shows the flights that will be managed next, with the currently selected aircraft highlighted. On the bottom of the screen, there is the schedule of requested actions for the different flights.

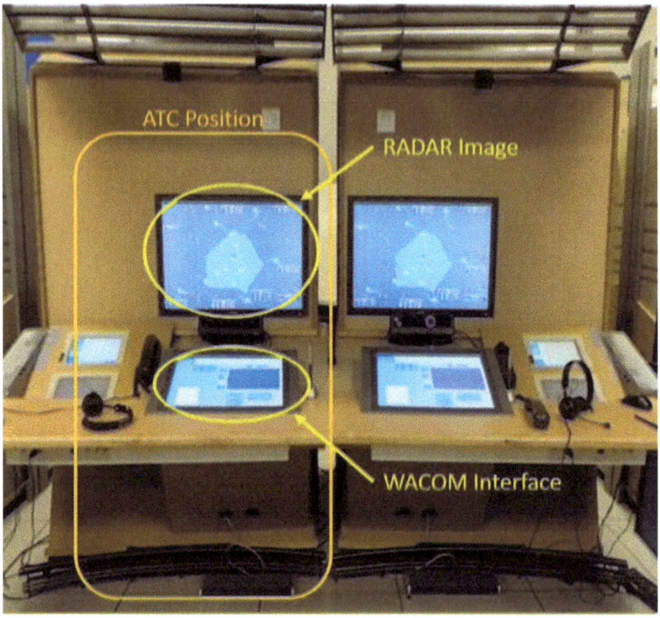


**Fig. 5.3** Operational position of the *en-route* ATCOs. It consisted in two screens. One of 30″ (RADAR Image) to display radar image, and a second of 21″ (WACOM Interface) to interact with the radar image (zoom, move, clearances, and information)

Furthermore, the Pseudo Pilots could modify the airplane upset depending on the ATCOs instructions or intentionally produce particular events (e.g., conflicts).

During the ATM simulation, the system recorded different parameters in order to define an index of the ATCO performance. In particular, the ATM simulator provided information on the reaction times and number of airplanes that the ATCO assumed in each specific task condition (Easy, Medium and Hard). The parameters considered as performance metric are the following:

- *Time Shoot*: the interaction time needed for delivering an aircraft to the next sector;
- *Time Assume*: the interaction time for accepting an aircraft in the sector;
- *Time Route*: the interaction time to display the graphic route of an aircraft;
- *Time Cancel*: the interaction time between triggering write recognition box and pressing of the cancel button;
- *Time Annul*: the interaction time between the opening of a pie menu or write recognition box and clicking on the radar image background;
- *Time Heading*: the interaction time between triggering the menu and validating the inserted value in the write recognition box;
- *Time Turn*: the interaction time between triggering the menu and the validation of the write recognition box;

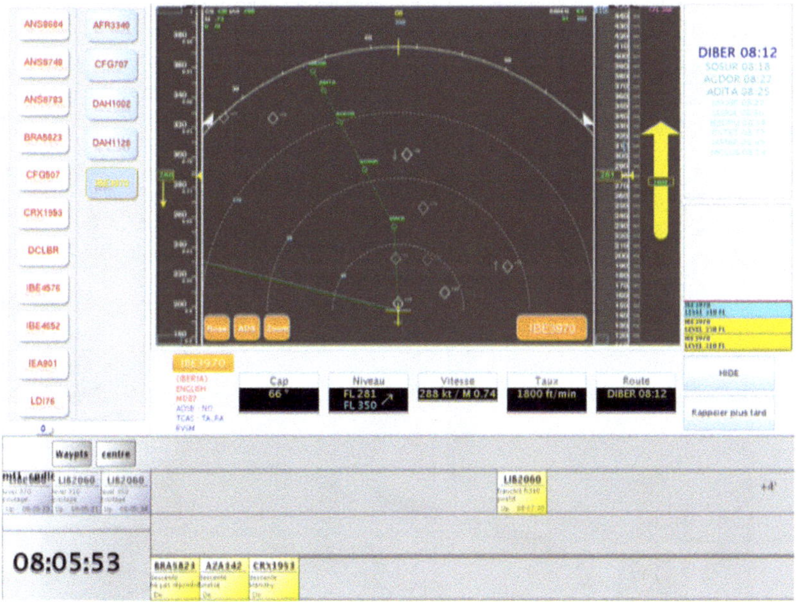


**Fig. 5.4** The interface of the Pseudo Pilots provided information about all the airplanes on the RADAR and those assumed by the ATCOs. Furthermore, the Pseudo Pilots could modify the airplane upset depending on the ATCOs instructions or intentionally to produce particular events (e.g., conflicts)

- *Time Flight Level*: the interaction time between triggering the menu and the validation of the write recognition box;
- *Time Direct*: the interaction time between triggering the menu and selecting the waypoint in the flight plan list.

Since each ATCO might adopt different strategies to manage air traffic, their *Reaction Times* (RT) have been weighted by taking into account the number of airplanes (nb) assumed in the different phases of the simulation. The performance index, *Weighted Mean Reaction Time* (WMRT), has then been defined as the average of the weighted RTs described previously.

## 5.4 Flight Simulators

The experiments which involved professional pilots have been performed in ecological settings by using full-motion flight simulators. In particular, the commercial pilots used the MD-80 (Figs. 5.5, 5.8) and AW189 (Fig. 5.7) simulator, whereas the military pilots used the MB-339 simulator (Figs. 5.6, 5.9). The simulators were builted with the same components and equipments of the aircrafts and helicopters,

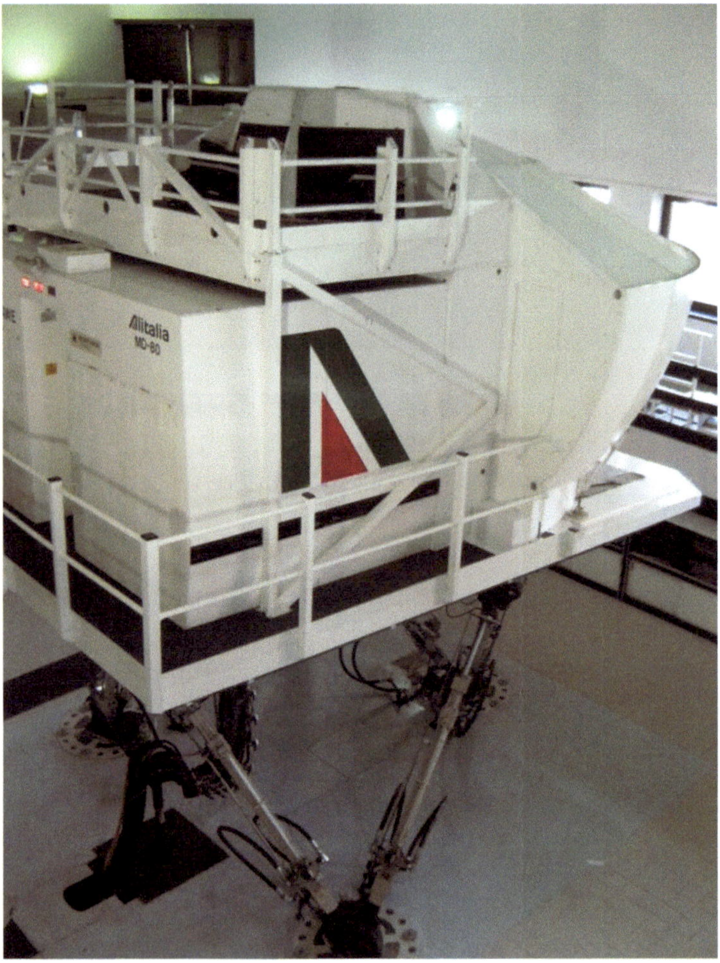


**Fig. 5.5** The flight simulations have been performed at the Training Center of the Alitalia airline in Fiumicino (Italy) by using the 6-axis full-motion MD-80 simulator and recruiting professional commercial pilots

and the environments were projected in *High Definition* (HD) within the cockpit or on a 180° curved screens.

The full-motion simulator had 6 axes and it could reproduce all the possible aircraft aptitudes and accelerations. The MD-80 simulator was hosted at the Training Center of the Alitalia airline in Fiumicino (Italy), the AW189 simulator was at the Training Academy of the AgustaWestland in Yeovil (UK), while the MB-339 simulator was hosted at the 61° Stormo of the Italian Air Force in Galatina (Italy).

The missions consisted of flight simulations during which unexpected events and failures were inserted during the taxi and after the cruise phases with aim to increase

**Fig. 5.6** The military flight simulations have been conducted at the Base of the 61° Stormo in Galatina (Italy) with the flight simulator of the MB339 airplane. The technology and components of the cockpit were the same of the jetfighter

**Fig. 5.7.** Rotary-wing pilots have worn the EEG cap during the entire simulations. The EEG portable systems have been placed on the back of the seats and no limitations to the operational activities have been produced

**Fig. 5.8** The commercial crew were asked to pilot the MD-80 within a mission from Rome to Pisa The pilot's brains activity were synchronized and then recorded along the entire mission

the mental workload demand and stress level of the mission. Specifically, the commercial mission was a flight from Rome Fiumicino (Italy) to Pisa (Italy) with a scheduled duration of 50 minutes. The weather conditions were roughly cloudy with possible windshake(W/S). In particular, at the departure airport (FCO), the pilots received information in order to presage the possibility of W/S (WIND 180V190 15G25) with temperature and high spread but without W/S alert. At the arrival airport (PSA), the weather conditions were WIND 130 15G20, BRK4500 and visibility of 8 km.

The flight simulations with the helicopter consisted of testing and comparing different avionic technologies by estimating the requested mental workload along the same mission. Specifically, the pilots had to takeoff, fly over the city, and then landing on a platform under different environmental visual conditions: *Good Visual Environment* (GVE), *Marginal Visual Environment* (MVE), and *Degraded Visual Environment* (DVE).

Instead, the military mission consisted of unusual attitudes recovery training. In particular, once reached the training area, the pilots have been distracted different times and they were asked to read some information in the *Flight Information Publication* (FLIP) book while the Examinator took the airplane to unusual attitudes (*Nose-Down*, *Nose-Up*, and *Spining*) from the remote control station. When distracted, the pilots had to realize that the aircraft's upset became unusual and then they had to recover from each of such conditions. In the experimental protocols with aircrafts, during the cruise phase, when the airplane reached the selected flight

**Fig. 5.9** The simulation environment has been projected on a 180° curved screen. The EEG system was placed on the back of the pilot's seat for avoiding operational limitations and the military pilots' brain activities were recorded for the entire simulation. The flight mission was monitored from the remote control station by Instructors

level (FL), the pilots were asked to execute a *Brain–Computer Interface* (BCI) task, which consisted of moving upward or downward a cursor on a computer screen by modulating the sensorimotor EEG rhythm ($\mu$-*rhythm*). The aim of this task was to hit the targets placed on the top and on the bottom of the screen: in particular, the target on the top represented the "point", and the target on the bottom represented the "line" of the International Morse code (Carron 1991). Successively, the BCI system encoded the targets into Morse symbols with the aim to generate the Morse message: *CQ SOS CU*. Such a message was a 30 balanced symbols sequence, 15 points and 15 lines, and it was based on a traditional radio communication code.

## References

Carron LP (1991) Morse code: the essential language. Amer Radio Relay League

Comstock JR (1994) MATB—Multi-Attribute Task Battery for human operator workload and strategic behavior research

Parasuraman R, Sheridan TB, Wickens CD (2000) A model for types and levels of human interaction with automation. IEEE Trans Syst Man Cybern-Part Syst Hum 30:286–297. doi:10.1109/3468.844354

# Chapter 6
# Neurophysiological Signals Processing

The objective of this chapter is to provide a general overview of the methods and technologies selected for the neurophysiological signal recording and analysis. In fact, acquiring signals and images (neuroimaging) from the human body has become vital for early diagnosis (Cartocci et al. 2015), marketing research (Di Flumeri et al. 2016b), human–machine interaction study (Rasmussen 1986; Zander and Kothe 2011), automation and system design (Giraudet et al. 2015; Parasuraman and Rizzo 2008). Such data can be in the form of electrobiological signals such as an *Electrocardiogram* (ECG) from the heart, *Electromyogram* (EMG) from muscles, *Electroencephalogram* (EEG) and *Magnetoencephalogram* (MEG) from the brain, *Electrogastrogram* (EGG) from stomach, and *Electrooculogram* (EOG) from eye nerves. Measurement can also have the form of one type of ultrasound or radiograph such as sonograph (or ultrasound image), *Computerized Tomography* (CT), *Magnetic Resonance Imaging* (MRI) or *functional MRI* (fMRI), *Positron Emission Tomography* (PET), and *Single Photon Emission Tomography* (SPET). Since the applications of fMRI, CT, PET are very limited in comparison with EEG or MEG (low-time resolution, limited and costly accessibility), we have finally chosen the EEG as technology for the brain signals recording.

## 6.1 EEG Analysis

The EEG has been recorded with a sampling frequency of 256 (Hz) by 64 electrodes referenced to both the earlobes and grounded to the AFz electrode. Simultaneously, the ECG electrode has been gathered from an electrode placed on the Erb's point. The EEG signals have then been digitally band-pass filtered by a fifth-order Butterworth filter [low-pass filter cut-off frequency: 30 (Hz), high-pass filter cut-off frequency: 1 (Hz)]. In order to remove eyeblink artifacts, the *Fpz* EEG channel has been used by using the regression-based algorithm REBLINCA



G. Borghini et al., *Industrial Neuroscience in Aviation*,
Biosystems & Biorobotics 18, DOI 10.1007/978-3-319-58598-7_6

(Di Flumeri et al. 2016a). With respect to other regressive algorithms (e.g., Gratton method, Gratton et al. 1983), the REBLINCA allows to preserve more EEG information in blink-free EEG epochs by using a specific threshold criterion that automatically recognize the occurrence of eyeblinks. In fact, only in this case the method corrects the contaminated EEG signals. If there is not any blink, the method will not perform any correction on the EEG epoch, avoiding the mutual contamination effect between EEG and EOG signals (Romero et al. 2008). Since the *Fpz* channel also contains EEG information, the application of traditional regression-based techniques may cause loss of important EEG information. To prevent such waste of information and, at the same time, to reject eyeblinks, a copy of the *Fpz* EEG channel (indicated in the following as *Fpz-EOG channel*) has been digitally band-pass filtered by a fifth-order Butterworth filter [low-pass filter cut-off frequency: 7 (Hz), high-pass filter cut-off frequency: 1 (Hz)]. This allowed to emphasize the eyeblinks component with respect to the brain activity. Then, by means of an automatic threshold criterion, based on a specific manipulation of the *Fpz* EEG channel itself, the time windows in which an eyeblink occured have been detected. Only in such windows, the filtered *Fpz-EOG* channel, properly multiplied for a specific weight estimated for each EEG channel during the regression phase, has been subtracted from all the EEG channels, similarly to the Gratton algorithm (Gratton et al. 1983). Despite a possible but very little loss of EEG information (just in correspondence of the eyeblink), the REBLINCA produced great improvements, in terms of system invasiveness and feasibility for real-time applications. This step has been performed because the eyeblinks contribution could affect the frequency bands related to the mental workload, in particular the theta EEG band. At this point, the EEG signal has then been segmented in epochs of 2 s, 0.125 s—shifted. These values have been chosen in order to have both a high number of observations in comparison with the number of variables, and in respect with the condition of stationarity of the EEG signal (Elul 1969). In fact, this hypothesis is necessary to proceed with the spectral analysis of the signal. For other artifacts different from the eyeblinks-related ones, specific procedures of the EEGLAB toolbox have been used (Delorme and Makeig 2004). In particular, three methods have been selected, that is *Threshold criterion*, *Trend estimation*, and *Sample-to-sample difference*. In the threshold criterion, the EEG epochs are marked as "artifact" if the EEG amplitude is higher than $\pm$ 100 ($\mu$V). In the trend estimation, the EEG epoch is interpolated in order to check the trend slope within the considered epoch. If such slope is higher than 10 ($\mu$V/epoch), the considered epoch will be marked as "artifact". The last check calculates the difference between consecutive EEG samples. If such difference, in terms of amplitude, is higher than 25 ($\mu$V), it means that an abrupt variation (no-physiological) happened, thus it will be marked as artifact. At the end, the EEG epochs marked as "artifact" have been removed with the aim to have an *artifact-free* EEG dataset from which estimating the brain parameters for the analyses. The percentage (with respect to the total number of epochs) of EEG epochs containing artifacts and removed from the EEG dataset was in average 20% (±13%). From the *artifact-free* EEG dataset, the *Power Spectral Density* (PSD) has been calculated for each EEG epoch using a of 2-second Hanning window (that means 0.5 (Hz) of frequency resolution). The application of a Hanning window helped to smooth the signal at the edges of the epochs, improving the accuracy of the PSD estimation (Harris 1978). Then, the EEG frequency bands have been defined for each participant by the estimation of the *Individual Alpha Frequency* (IAF) value (Klimesch 1999). In order to have a precise estimation of the alpha peak and, hence of the IAF, the participants have been asked to keep the eyes closed for a minute before starting with the experiments. Finally, a spectral features matrix (EEG channels x Frequency bins) has been obtained in the considered frequency bands. In particular, only the theta (IAF-6 ÷ IAF-2) and alpha (IAF-2 ÷ IAF + 2) rhythms have been considered as parameters for the mental states assessment.

## 6.2 Estimation of Brain Cortical Sources Current Density

Cortical activity has been estimated from EEG scalp recordings by employing the high-resolution EEG technologies (Astolfi et al. 2007; Babiloni et al. 2005; De Vico Fallani et al. 2010) with the use of the average head model from McGill University (Ding et al. 2005). Electrode positions over the scalp have been obtained individually for each participant by using the *Photomodeler software* (Eos System Inc). The cortical model consisted of about 8,000 dipoles uniformly disposed on the cortical surface. The estimation of the current density strength for each dipole has been obtained by solving the *electromagnetic linear inverse problem* (He et al. 2006; Nunez and Cutillo 1995). In particular, the solution of the linear system,

$$\mathbf{Ax} = \mathbf{b} + \mathbf{n} \tag{6.2.1}$$

at a particular time instant $t$ provides an estimation of the dipole source configuration $\boldsymbol{x}$ at time $t$ that generates the measured EEG potential distribution $\mathbf{b}$ in the same instant. The system also includes the measurement noise $\mathbf{n}$, assumed to be normally distributed. $\mathbf{A}$ is the lead field matrix, where each $j$th column describes the potential distribution generated on the scalp electrodes by the $j$th unitary dipole. The current density solution vector of Eq. (6.2.1) has been obtained as

$$\xi = \arg\min_{x}\left( \|\mathbf{A}x - b\|_{\mathbf{M}}^{2} + \lambda^{2}\|x\|_{\mathbf{N}}^{2} \right) \tag{6.2.2}$$

where $\mathbf{M}$, $\mathbf{N}$ are the matrices associated to the metrics of the data and of the source space, respectively, $\lambda$ is the regularization parameter and $\| \ x \ \|_N$ represents the $N$ norm of the vector $x$ (Babiloni et al. 2005; Grave de Peralta Menendez and Andino 1999). The solution of Eq. (6.2.2) is given by the inverse operator $\boldsymbol{G}$:

$$\xi(t) = \mathbf{Gb}(t), \tag{6.2.3}$$

where

$$\mathbf{G} = \mathbf{N}^{-1}\mathbf{A}'\left(\mathbf{AN}^{-1}\mathbf{A}' + \lambda\mathbf{M}^{-1}\right)^{-1} \tag{6.2.4}$$

The optimal determination of the regularization term $\lambda$ of this linear system has been obtained by the *L-curve approach* (Hansen 1992). As a metric in the data space, the identity matrix has been used, while as a norm in the source space, the following metric was adopted:

$$\left(\mathbf{N}^{-1}\right)_{ii} = \|\mathbf{A}_{\cdot i}\|^{-2} \tag{6.2.5}$$

where $(\mathbf{N}^{-1})_{ii}$ is the $i$th element of the inverse of the diagonal matrix $\mathbf{N}$ and all the other matrix elements $\mathbf{N}_{ij}$ are set to 0. The $L_2$ norm of the $i$th column of the lead field matrix $\mathbf{A}$ is denoted by $\|\mathbf{A}_i\|$. Using the relations described above, an estimation of the signed magnitude of the dipolar moment for each cortical dipole has been obtained for each time point $t$. As the orientation of the dipole has been defined to be perpendicular to the local cortical surface in the head model, the estimation process returned a scalar rather than a vector field. The spatial average of the signed magnitude, of all the dipoles belonging to a particular ROI at each time sample was used to estimate the waveforms of cortical ROI activity in that ROI, indicated as $\boldsymbol{\rho}(t)$ to highlight their time dependence. Spatial averaging can be expressed in terms of matrix multiplication by a matrix $\mathbf{T}$. This matrix is sparse and has as many rows as ROIs, and as many columns as the number of dipole sources.

ROI cortical current density waveforms can then be expressed as

$$\boldsymbol{\rho}(t) = \mathbf{Tx}(t) = \mathbf{TGb}(t) = \mathbf{G}_{\mathrm{ROI}}\mathbf{b}(t), \quad \text{where } \mathbf{G}_{\mathrm{ROI}} = \mathbf{TG} \tag{6.2.6}$$

where $\mathbf{b}(t)$ is the array of the waveforms recorded from the scalp electrodes and $\mathbf{x}(t)$ is the array of the cortical current density waveforms estimated at the cortical surface. The $\mathbf{G}_{\mathrm{ROI}}$ matrix only depends on geometrical factors, and can thus be computed and stored offline. The matrix multiplication can be interpreted as a spatial filtering of the scalp potential $\mathbf{b}(t)$, using the elements of $\mathbf{G}_{\mathrm{ROI}}$ as weights. In this way, we could obtain time-varying waveforms at the level of different cortical areas. In the next paragraph, such cortical areas will be described as coincident with particular Brodmann areas for all the participants involved in the present study.

## 6.3 Generation of the Regions of Interest (ROIs)

Brain cortical *Regions of Interest* (ROIs) have been considered to investigate brain activations at the cortex level. The ROIs were segmented on the basis of the *Brodmann areas* (BA) classification (Brodmann's 2006), and the considered ROIs were those suggested by literature evidences related to decision-making, information processing, decision planning. In particular, ROIs within the *dorsolateral prefrontal cortical* (DLPFC), *Anterior Cingulate Cortex* (ACC), *Cingulate Motor Area* (CMA), and parietal regions have been selected: BA5, BA6, BA7, BA8, BA9, BA10, BA19, BA24, BA25, BA33, and BA40.

## 6.4 EEG-Based Mental Workload Index ($W_{\mathrm{EEG}}$)

The asSWLDA classifier (see Sect. 6.7.1 for detailed description) has been used to select within the calibration dataset the most relevant EEG spectral features to discriminate the variations of the considered mental status (e.g., mental workload). Figure 6.1 shows the algorithm steps for the *EEG-based workload index* definition. The asSWLDA classifier assigned to each relevant feature specific weights ($w_{i\,\mathrm{train}}$), plus an intercept ($b_{\mathrm{train}}$). On the contrary, weights related to not relevant features for the classification model have been set to 0. This step represented the calibration phase of the algorithm. The parameters estimated during the calibration phase have been then used to calculate the *Linear Discriminant Function* ($y_{\mathrm{test}}(t)$) over the testing EEG dataset (testing phase, Eq. 6.4.1), defined as the linear combination of

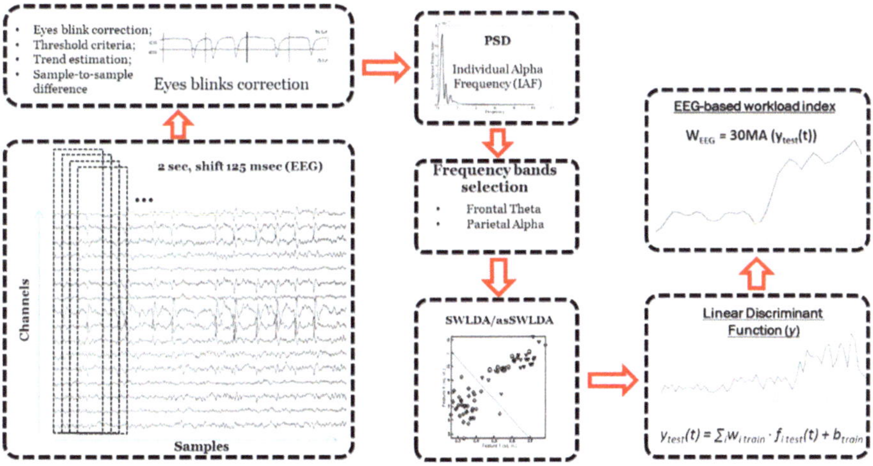


**Fig. 6.1.** EEG-based workload index estimation. The filtered (1–30 (Hz)) EEG has been segmented into epochs of 2 (s), shifted of 0.125 (s), and the filtered (1–7 (Hz)) EOG signal has been used to remove the eyes-artifact contribution. Other sources of artifacts have been deleted by using specific algorithms. Then, the PSD has been evaluated for each epoch ($\mathrm{Epoch}_{\mathrm{PSD}}$) taking into account only the EEG frequency bands and channels correlated with the mental workload (frontal theta and parietal alpha bands). The asSWLDA has then been used to select the most relevant brain spectral features for the discrimination of the mental workload levels. The linear discriminant function has been calculated on the EEG testing dataset, and the $W_{\mathrm{EEG}}$ has been defined as the moving average of 30 s applied to the linear discriminant function ($y_{\mathrm{test}}(t)$)

the testing spectral features ($f_{i\ \text{test}}$) and the classifier weights ($w_{i\,\text{train}}$), plus the bias ($b_{\text{train}}$). Finally, a moving average of $k$ seconds (kMA) has been applied to the $y_{\text{test}}(t)$ function in order to smooth it out by reducing the variance of the measures, and the result has been named *EEG-based Workload Index* ($W_{\text{EEG}}$, Eq. 6.4.2). The higher is the $k$ value, the less will be the variance of the measure. Accordingly with Subject Matter Experts (SMEs), for a proper evaluation of the investigated mental status the $k$ value has been set to 30 s.

$$y_{\text{test}}(t) = \sum\nolimits_i w_{i_{\text{train}}} * f_{i_{\text{test}}}(t) + b_{\text{train}} \quad (6.4.1)$$

$$W_{\text{EEG}} = kMA(y_{\text{test}}(t)), \quad k = 30\ (\text{s}) \quad (6.4.2)$$

To evaluate the reliability of the classifier in discriminating the mental states levels, the *Area Under Curve* (AUC) values of the *Receiver Operating Characteristic* (ROC, Bamber 1975) has been calculated by considering a couple of experimental conditions $W_{\text{EEG}}$ distributions (e.g., Easy vs. Hard).

## 6.5 Inter-brain Connections

The first EEG hyperscanning study was reported by Babiloni et al. (2006) and involved teams of four participants playing *Tressette*, an Italian card game. Since then, there have been several works claiming cognitive coupling between people engaged in social interaction (Astolfi et al. 2011; Babiloni et al. 2006, 2012; Burgess 2013a; De Vico Fallani et al. 2010; Dumas et al. 2010; Faes and Nollo 2011; Fallani et al. 2012; Kawasaki et al. 2013; Kong et al. 2017; Naeem et al. 2012; Sakkalis 2011; Tognoli 2008). The methods used to establish neural coupling between people have been very consistent and nearly all studies have used one of the following methods: (i) covariance in EEG amplitude or power (French and Beaumont 1984), (ii) coherence (Baccalá and Sameshima 2001) such as the *Partial Directed Coherence* (PDC), and (iii) phase synchrony (Lachaux et al. 1999), mostly the *Phase Locking Value* (PLV).

In most cases, even if covariances in EEG amplitudes or power (PSD) have been found, this measure is a weak form of association, and although it is suggestive of neural coupling, it is by no means conclusive (Burgess 2013a).

### *6.5.1 Coherence (COH)*

Coherence is a mathematical technique that quantifies the frequency and amplitude of the synchronicity of neuronal patterns of oscillating brain activity (Bezerianos et al. 2015; French and Beaumont 1984). This technique quantifies the neuronal patterns of synchronicity measured between spatially separated scalp electrodes

(EEG) or coils (MEG). Coherence is an estimate of the consistency of relative amplitude and phase between signals detected in coils or electrodes within a set frequency band. In sensor space if signals are in phase then their amplitudes will add, if they are out of phase the signals will subtract possibly reducing the coherence value. In source space, the amplitude of the underlying source can be used to determine the strength of the connectivity. In most cases, the coherence is estimated by the following formula:

$$\mathrm{COH}_{xy} = \frac{\left|\frac{1}{N}\sum_{k=1}^{N} Y_k(f)X_k^*(f)\right|}{\sqrt{\frac{1}{N}\sum_{k=1}^{N} {}_k(f)X_k^*(f)X \cdot \frac{1}{N}\sum_{k=1}^{N} Y_k(f)Y_k^*(f)}} \tag{6.5.1}$$

where $X_k(f)$ denotes FFT of the $k$th segment of the time series $x(t)$ at frequency $f$ and * indicates the transpose and complex conjugate.

### 6.5.2 *Partial Directed Coherence (PDC)*

The PDC was the approach used in the very first EEG hyperscanning study (Babiloni et al. 2006), and in the work described in this book. PDC is based on multivariate autoregressive modeling and Granger Causality (Granger 2004; Kong et al. 2015; Taya et al. 2015) and it is designed to be able to show the direction of information flow (linear) between two systems (Baccalá and Sameshima 2001). As such, PDC seems ideally suited to role of identifying inter-brain coupling in hyperscanning studies, at least in those cases where when one person's behavior is driving another's. The PDC from a time series $y(t)$ to a time series $x(t)$ is defined as

$$\mathrm{PDC}_{xy}(f) = \frac{A_{xy}(f)}{\sqrt{a_y^*(f) \cdot a_x(f)}} \tag{6.5.2.}$$

where $A_{xy}(f)$ is an element in $\boldsymbol{A}(f)$ which is the Fourier Transform of the multivariate autoregressive (MVAR) model coefficients, $\boldsymbol{A}(t)$, of the time series; $a_y(f)$ is $y$th column of $\boldsymbol{A}(f)$.

### 6.5.3 *Phase Locking Value (PLV)*

The third class of measures of brain-to-brain coupling consists of evaluating phase synchrony (Dumas et al. 2012; Lindenberger et al. 2009; Sänger et al. 2013). The first use of phase synchronization as a measure of coupling with electrophysiological data was made by Tass et al. (1998), who defined synchronization as occurring when $|\varphi_{n,m}| < \mathrm{const}$ where const is a suitably constant value, $n$ and $m$ are integers, $\varphi_{n,m}(t)$ is the phase difference $(n\phi_1(t) - m\phi_2(t))$, and $\phi_{1,2}$ are the phases of the two oscillators. The most widely used index of phase locking adopted in hyperscanning studies is the *Phase Locking Value* (PLV), which is a well-suited measure for capturing the rapid flow of information between people in social situations (Lachaux et al. 1999). The PLV, as originally defined by Lachaux et al. (1999), is estimated by

$$\mathrm{PVL}_n = \frac{1}{N}\left|\sum_{k=1}^{N} e^{-i(\phi_{(t,k)} - \psi_{(t,k)})}\right| \quad (6.5.3)$$

where $N$ is the number of trials, $\phi_{(t,k)}$ is the phase on trial $k$ at time $t$ in the channel $\phi$, and $\psi_{(t,k)}$ in channel the $\psi$. The $\mathrm{PLV}_n$ varies between 0 and 1 where 1 indicates perfect phase locking, and 0 indicates no phase locking.

## 6.6 Autonomic Data Analysis

The *Heart Rate* (HR) and the *Eyeblink Rat*e (EBR) have been estimated by calculating the distance between consecutive peaks occurring in the ECG and in the EOG signals. In particular, the R-peak and the eyeblink-peak have been used for the HR and EBR estimation, respectively. The ECG and EOG signals have been band-pass filtered, and then the eyeblink and R-peak were found by applying a threshold criterion to the raw signals estimated individually for each participant. The peaks were verified in terms of time-duration, amplitude and time-distance between consecutive peaks, to rejects those due to artifacts. The EBR and HR were estimated by the following formula:

$$\mathrm{Rate} = \left(\frac{\mathrm{fs}}{\mathrm{ibi}}\right) * 60 \quad (6.6.1)$$

where fs is the sample frequency and the ibi (inter-beat-interval) is the distance between two consecutive peaks. As reported by Kaber et al. (Kaber et al. 2007), many researches have explored the use of HR for assessing operators' workload in real time and to provide the basis for dynamic tasks' allocation—to automation or manual control (Scerbo et al. 2001; Wilson 2002; Wilson and Russell 2003). These studies have revealed that an increased HR could be related to an increased mental workload: as the *Central Nervous System* (CNS) controls HR by varying the impulse traffic in the sympathetic and parasympathetic nerve fibers terminating in the sinoatrial node (Cherkas et al. 2015; Gordan et al. 2015; White and Raven 2014) an *HR index* was defined as the linear combination of the HR and the arterial pressure (R-peak's amplitude–HRamp). On the contrary, the EBR has been demonstrated to be inversely correlated with the increase of the mental workload (Lal and Craig 2002). The results proposed in the following sections will then be represented as *r-square* values. The *Coefficient of Determination* ($r^2$), or *r-square*'s value is a statistical measure computed over a pair of sample distributions, giving a measure of how strongly the means of the two distributions differ in relation to variance (Steel et al. 1996). For two datasets $x$ and $y$, the *r-square* is calculated by the following formula:

$$r^2 = \frac{\text{cov}(x, y)^2}{\text{var}(x)\text{var}(y)} \tag{6.6.2}$$

where cov($x$, $y$) is the covariance between the two datasets, and var(•) is the variance of the single dataset. As $0 < r^2 < 1$, by definition, a *signed r-square* has been derived by multiplying the coefficient of determination by the sign of the slope of the corresponding linear model of the regression analysis. In this way, it has been possible to obtain information not only about if the two datasets were different, but also about the direction of such difference. Therefore, if the $r$-square value is zero, it means that there are no differences between the considered experimental conditions.

## 6.7 Machine Learning

Machine learning is a discipline focused on two interrelated questions: "*How can one construct computer systems that automatically improve through experience*?" and "*What are the fundamental statistical computational-information-theoretic laws that govern learning systems*?" The study of machine learning is important, both for addressing these fundamental scientific and engineering questions, and for the highly practical computer software that has been produced and used in many application fields (Jordan and Mitchell 2015). Machine learning has progressed dramatically over the past two decades, from just a laboratory curiosity to a real practical technology, until to a widespread commercial use. Within artificial intelligence (AI), machine learning has emerged as the method of choice for developing practical software for computer vision, speech recognition, natural language processing, robot control, and other applications. Many developers of AI systems now recognize that, for many applications, it can be far easier to calibrate a system by showing it examples of desired *input–output* behavior than to program it manually by anticipating the desired response for all possible inputs. The effect of machine learning has also been felt broadly across computer science and across a range of industries concerned with big-data issues, such as consumer services, the diagnosis of faults in complex systems, and the control of logistics chains. There has been a similarly broad range of effects across empirical sciences, from biology to cosmology to social science, as machine learning methods have been developed to analyze high throughput experimental data in novel ways. In particular, under the recognition of the user's mental states (e.g., mental workload), such techniques should be able to extract the most significant characteristics closely related to the examined mental state from the big amount of physiological data, that can be used to assess the examined user's mental state on different working days.

A diverse array of machine learning algorithms has been developed to cover the wide variety of data and problem types exhibited across different machine learning problems (Murphy et al. 2012). Conceptually, machine learning algorithms can be viewed as searching through a large space of candidate programs, guided by training experience, to find a program that optimizes the performance metric. Machine learning algorithms vary greatly, in part by the way in which they represent candidate programs (e.g., decision trees, mathematical functions, and general programming languages) and in part by the way in which they search through this space of programs (e.g., optimization of algorithms with well-understood convergence guarantees and evolutionary search methods that evaluate successive generations of randomly mutated programs).

Many algorithms focus on function approximation problems, where the task is embodied in a function (e.g., given an input transaction, output a "fraud" or "not fraud" label), and the learning problem is to improve the accuracy of that function, with experience consisting of a sample of known input–output pairs of the function. In some cases, the function is represented explicitly as a parameterized functional form; in other cases, the function is implicit and obtained via a search process, a factorization, an optimization procedure, or a simulation-based procedure. Even when implicit, the function generally depends on parameters or other tunable degrees of freedom, and training corresponds to finding values for these parameters that optimize the performance metric. Whatever the learning algorithm, a key scientific and practical goal is to theoretically characterize the capabilities of specific learning algorithms and the inherent difficulty of any given learning problem: "*How accurately can the algorithm learn from a particular type and volume of training data? How robust is the algorithm to errors in its modeling assumptions or to errors in the training data? Given a learning problem with a given volume of training data, is it possible to design a successful algorithm or is this learning problem fundamentally intractable*?" Such theoretical characterizations of machine learning algorithms and problems typically make use of the familiar frameworks of statistical decision theory and computational complexity theory. In fact, attempts to characterize machine learning algorithms have led—theoretically—to blends of statistical and computational theories in which the goal is to characterize simultaneously the sample complexity (how much data are required to learn accurately) and the computational complexity (how much computation is required) and to specify how these depend on features of the learning algorithm such as the representation it uses for what it learns (Chandrasekaran and Jordan 2013; Decatur et al. 2000; Shalev-Shwartz and Zhang 2013). As a field of study, machine learning sits at the crossroads of computer science, statistics and a variety of other disciplines concerned with automatic improvement over time, and inference and decision-making under uncertainty. Related disciplines include the psychological study of human learning, the study of evolution, adaptive control theory, the study of educational practices, neuroscience, organizational behavior, and economics. Although the past decade has seen increased crosstalk with these other fields, we are just beginning to tap the potential synergies and the diversity of formalisms and experimental methods used across these multiple fields for studying systems, which increasingly improve alongside experience.

### *6.7.1 Supervised Learning Systems*

The most widely used machine learning methods are the supervised learning ones. Supervised learning systems, including spam classifiers of e-mails, face recognizers over images and medical diagnosis systems for patients, all exemplify the function approximation problem where the training data take the form of a collection of $(x, y)$ pairs and the goal is to produce a prediction $y*$ in response to a query $x*$. The inputs $x$ may be classical vectors, or they may be more complex objects such as documents, images, DNA sequences, or graphs. Similarly, many different kinds of output $y$ have been studied. Much progress has been made by focusing on the simple binary classification problem in which $y$ takes on one of two values (for example, "spam" or "not spam"), but there has also been abundant research on problems such as multiclass classification, where $y$ takes on one of $K$ labels. Supervised learning systems generally form their predictions via a learned mapping $f(x)$, which produces an output $y$ for each input $x$ (or a probability distribution over $y$ given $x$). Many different forms of mapping $f$ exist, including decision trees, decision forests, *linear regression* (LDA), *support vector machines* (SVM), *neural networks* (NN), kernel machines, and Bayesian classifiers.

### *6.7.2 Unsupervised Learning Systems*

While much of the practical success in deep learning has come from supervised learning methods, efforts have also been made to develop deep learning algorithms that discover useful representations of the input without the need for labeled training data (Hinton and Salakhutdinov 2006). The general problem is referred to as unsupervised learning, a second paradigm in machine learning research (Murphy et al. 2012). Broadly, unsupervised learning generally involves the analysis of unlabelled data, under assumptions of certain structural properties of the data (e.g., algebraic, combinatorial, or probabilistic). For example, one can assume that data lie on a low-dimensional manifold and aims to identify that manifold explicitly from the data itself. Dimension reduction methods, including principal components analysis, manifold learning, factor analysis, random projections, and auto-encoders (Murphy et al. 2012) make different specific assumptions regarding the underlying manifold (e.g., that it is a linear subspace, a smooth nonlinear manifold, or a collection of submanifolds).

### 6.7.3 Reinforcement Learning System

A third major machine learning paradigm is the reinforcement learning (Mnih et al. 2016; Montague 1999). Here, the information available in the training data is intermediate between the supervised and unsupervised learning. Instead of training examples that indicate the correct output for a given input, the training data in reinforcement learning are assumed to provide only an indication as to whether an action is correct or not; if an action is incorrect, there remains the problem of finding the correct action. More generally, in the setting of sequences of inputs, it is assumed that reward signals refer to the entire sequence; the assignment of credit or blame to individual actions in the sequence is not directly provided.

Although these three learning paradigms help to organize ideas, much current research involves blending across these categories. For example, *semi-supervised* learning makes use of unlabeled data to augment the latter in a supervised learning context, and discriminative training blends architectures developed for unsupervised learning with optimization formulations that make use of labels.

Machine learning model selection does not mean only to choose the model that better fits the calibrating data, but also to select from a family of models. Many issues influence the design of learning algorithms across all of the previous paradigms, including whether data are available in batches or arrive sequentially over time, how data have been sampled, requirements that learned models be interpretable by users, and robustness issues that arise when data do not fit prior modeling assumptions. Additionally, from a neurophysiological perspective, different cognitive processes and mental states may result in different patterns of brain activations. Machine learning algorithms are seen as pattern recognition techniques that classify each pattern into a class according to its features. The features are measured or derived from the properties of the signals which contain the discriminative information needed to distinguish their different types. The design of a suitable set of features is a challenging issue. The information of interest in brain signals is hidden in a highly noisy environment, and brain signals comprise a large number of simultaneous sources. A signal that may be of interest could be overlapped by multiple signals depending on different brain processes or artifacts. Therefore, one of the critical point of machine learning techniques is the selection of the relevant features within the available ones. High-dimensional feature vectors are not desirable due to the “curse of dimensionality” in calibrating the algorithms. In other words, the feature selection may be attempted examining all possible subsets of the features, thus it might take a long time and computational demand. However, the number of possibilities grows exponentially, making an exhaustive search impractical for even a moderate number of features.

The aim of BCI applications is the recognition n of a user’s intentions on the basis of a feature vector that characterizes the brain activity provided by the feature dataset. Either regression or classification algorithms can be used to achieve this goal (Lotte et al. 2007). Regression algorithms employ the features extracted from

EEG signals as independent variables to predict user intentions. In contrast, classification algorithms use the features extracted as independent variables to define boundaries between the different classes in feature space. McFarland and Wolpaw (2005) illustrated the differences between the two alternatives. For a two-data classes case, both the regression and the classification approach require the parameters of a single function to be determined. In a four-data classes case, assuming that the targets are distributed linearly, the regression approach still requires only a single function. In contrast, the classification approach requires the determination of three functions, one for each of the three boundaries between the four classes. Therefore, the classification approach might be more useful for two-target applications and the regression approach may be preferable for greater numbers of targets, when these targets can be ordered along one or more dimensions. Classification algorithms can be developed via either offline, online', or both of conditions (Nicolas-Alonso and Gomez-Gil 2012). The offline session involves the examination of dataset. The statistics of the data may be estimated from observations across entire sessions and long-term computations may be performed. The analyst, with the aim of fine-tuning the algorithms, can then review the results. In contrast, online sessions provide a means of evaluation in a real-world environment. The data are processed in a causal manner and the algorithms are tested in an environment in which the users change over the time as a result. Although some researchers test new algorithms with only offline data, both offline simulation and online (or simulated online) experiments are necessary for effective algorithm design in closed-loop systems. In other words, offline simulation and cross-validation can be valuable methods to develop and test new algorithms, but only online (or simulated online) analysis can yield solid evidence of machine learning system performance (Daly and Wolpaw 2008; McFarland et al. 2006; Townsend et al. 2010). Traditional classification algorithms have been calibrated through labeled dataset. It is assumed that the classifier is able to detect the patterns of the brain signal recorded in online sessions with feedback. The design of the classification step involves the choice of one or several classification algorithms from many alternatives. Linear and nonlinear classification algorithms have been proposed in the literature, such as *k*-nearest neighbor classifiers, linear classifiers, support vector machines, and neural networks, among others (Table 6.1). Linear methods assume a linear separability of the data classes. The separability of the classes is measured by two quantities: how far are the projected class means apart (should be large), and how big the variance of the data in this direction is (should be small). This can be achieved by maximizing the so-called *Rayleigh coefficient* of between and within class variance (Fisher 1936; Fukunaga 1990). On the contrary, when the classes are not separable linearly, or the source of the data to be classified is not well understood, methods for finding good nonlinear transformations are required. For example, the classification in input space requires some complicated nonlinear (multiparameter) ellipsoid classifier, and appropriate feature space representation.

**Table 6.1.** Summary of the most popular machine learning techniques

| Generative model | Approach | Properties |
|---|---|---|
| Linear | Bayesian analysis | – It assigns the observed feature vector to the labeled class which it has the highest probability of belonging to<br>– It produces nonlinear decision boundaries |
| | LDA | – Simple classifier with acceptable accuracy<br>– It fails in presence of outliers or strong noise. Regularization required<br>– Usually two class. Exits multiclass version |
| | SWLDA | – Extension of the LDA that performs feature space reduction by selecting suitable features to be included in the discriminant analysis<br>– Combination of forward and backward LDA<br>– Multi—classes analysis |
| NonLinear | SVM | – Linear and nonlinear (Gaussian) modalities<br>– Binary or multiclass method<br>– It maximizes the distance between the nearest training samples and the hyperplanes<br>– It fails in the presence of outliers or strong noise |
| | k—NNC | – It uses metric distances between the test feature and their neighbors<br>– Multiclasses<br>– Efficient with low-dimensional feature vectors<br>– Very sensitive to the dimensionality of the feature vectors |
| | ANN | – Very flexible classifier<br>– Multiclass<br>– Multiple architectures (PNN, Fuzzy ARTMAP ANN, FIRNN, PeGNC) |

Finally, certain inherent dangers of classification algorithm usage should be pointed out. Although classification algorithms have clearly helped to characterize task-relevant brain states, several pitfalls may occur when these algorithms are used. Bias and variance of the estimated error of the algorithms, and their overfitting are the main source of problems (Lemm et al. 2011). If a classifier is overfitted, then it will only be able to classify classes in calibration or similar data. Classification error is estimated by means of cross-validation. Once a classification algorithm is calibrated, the algorithm is validated on a testing dataset, which should be independent on the calibration dataset. This procedure is usually repeated several times, using different partitions of the calibration data. The resulting validation errors are averaged across multiple rounds (cross-validations). This approach presents some inherent dangers that must be prevented, because some elements of the partition may not be independent of each other (*auto-validation*) or may not be identically distributed, among other reasons (Lemm et al. 2011).

### 6.7.4 Linear Discriminant Analysis (LDA)

Linear Discriminant Analysis (LDA) is usually applied to classify patterns into two classes, although it is possible to extend this method to multiple classes (Garrett et al. 2003). For a two-class problem, LDA assumes that the classes are linearly separable. According to this assumption, LDA defines a linear discrimination function $y(x)$ which represents a hyperplane in the feature space in order to distinguish the classes. The class which the feature vector belongs to, will depend on the side of the plane where the vector is found. In the case of an N-class problem ($N > 2$), several hyperplanes are used.

The decision plane can be represented mathematically as

$$y(\boldsymbol{x}) = w^T\boldsymbol{x} + w_0 \tag{6.7.1}$$

where $w$ is known as the weight vector, $\mathbf{x}$ is the input feature vector, and $w_0$ is a bias. The input feature vector is assigned to one class or the other on the basis of the sign of $y(\mathbf{x})$.

There are many methods to compute $w$. For example, $w$ may be calculated as (Fisher 1936)

$$w = \sum_c^{-1} (\mu_2 - \mu_1) \tag{6.7.2}$$

where $\mu_i$ is the estimated mean of the class $i$, and $\sum_c = \frac{1}{2}(\sum_1 + \sum_2)$ is the estimated common covariance matrix. The estimators of the covariance matrix and of the mean are calculated as

$$\sum c = \frac{1}{n-1}\sum\nolimits_{i=1}^{n} (x_i - \mu)(x_i - \mu)^T \tag{6.7.3}$$

$$\mu = \frac{1}{n}\sum_{i=1}^{n} x_i \tag{6.7.4}$$

where $\mathbf{x}$ is a matrix containing $n$ feature vectors $x_1, x_2, \ldots, x_n \in \mathbb{R}^d$. The estimation of the covariance defined in Eq. (6.7.3) is unbiased and has good properties under usual conditions. Nevertheless, it may become imprecise in some cases where the dimensionality of the features is too high in comparison with the number of available trials. The estimated covariance matrix is different from the true covariance matrix, because the large eigenvalues of the original covariance matrix are overestimated and the small eigenvalues are underestimated. It leads to a systematic error, which degrades LDA performance (Blankertz et al. 2010). For this reason, a new procedure has been proposed to estimate the covariance, improving the standard estimator defined in the Eq. (6.7.3). The new standard estimator of the covariance matrix is given by

$$\sum_c (\gamma) = (1-\gamma)\sum_c + \gamma v I \tag{6.7.5}$$

The $\gamma$ value is referred to as a shrinkage parameter and is tunable between 0 and 1. $v$ is defined as trace$(\sum_c)/d$ with $d$ being the dimensionality of the features space. The selection of a shrinkage parameter implies a trade-off and is estimated on the basis of the input data (Vidaurre et al. 2009). Some improved algorithms have been introduced based on LDA (Hoffmann et al. 2008), such as *Fisher LDA* (FLDA) and *Bayesian LDA* (BLDA). In the first example, performance was improved by projecting the data to a lower dimensional space, in order to achieve larger intervals between the projected classes and, simultaneously, to reduce the variability of the data in each class. However, FLDA does not work well when the number of features becomes too large in relation to the number of calibrating examples. This is known as 'the small sample size problem' (Hoffmann et al. 2008). The second modification can be seen as an extension of the FLDA: BLDA solves the small sample size problem by introducing a statistical method known as *regularization*, which is estimated through a Bayesian analysis of calibrating data and is used to prevent *overfitting* of high-dimensional and possibly noisy datasets. In comparison to FLDA, the BLDA algorithm provides higher classification accuracy and bit-rates, especially in those cases where the number of features is large (Hoffmann et al. 2008). Additionally, BLDA requires only slightly more computation time, which is a crucial requirement in real BCI systems.

### 6.7.5 Stepwise Linear Discriminant Analysis (SWLDA)

The SWLDA is an extension of the FLDA, which performs a reduction of the features' space by selecting the most significant ones. In particular, the SWLDA consists of a combination of forward selection and backward elimination analyses, where the input features are weighted using ordinary least-squares regression to predict the target class label. The method starts by creating an initial model of the discriminant function with the most statistically significant feature for predicting the target labels (e.g., $p$-value < 0.5). Then, at each new step, a new term is added to the model, and a backward stepwise analysis is performed to remove the least statistically significant feature (e.g., $p$-value > 0.1). This process goes on until the predefined number of significant features is reached or until there are no features satisfying the entry/removal condition (Draper 1998a).

Regarding the mental workload evaluation, it has been demonstrated that linear classifiers outperformed the nonlinear ones (Craven et al. 2006; Krusienski et al. 2006). In particular, hereby the SWLDA has been chosen since it was proved to be one of the best classifiers for the real-time mental workload evaluation (Berka et al. 2007; Rabbi et al. 2009). Generally, several cognitive processes interact with each other during the execution of a task, and since one of our research aims was to track only the variation of a specific phenomenon (i.e., mental workload or training), the SWLDA has been selected because of the possibility to set its parameters—hence statistic criteria, for the features' selection (see paragraph 6.7.1 for more details). In fact, the SWLDA is able to weight the contribution of the features to the classification model, and to find out if each feature has to be inserted into or removed from the model. As a consequence, by properly acting on the SWLDA's features selection parameters (see paragraph 6.7.1), if the considered phenomenon is sufficiently emphasized by the task, it will be possible to hook and track the variation of such phenomenon ongoing the task execution. In fact, the SWLDA will consider (into the model) only those brain features able to describe univocally the examined cognitive phenomenon, and which will be always turned up by the execution of the proposed task, therefore stable over time.

### 6.7.6 Automatic-Stop SWLDA (asSWLDA)

Nowadays, the effects of day-to-day fluctuations in the operator's physiology have not been thoroughly assessed while operators are engaged in complex tasks. Different studies showed that the classifiers' performance, in evaluating the user's different mental workload levels, dramatically decreases over days (Burgess and Gruzelier 1993; Christensen et al. 2012; McEvoy et al. 2000; Salinsky et al. 1991). Recent attempts have been made to test the stability of different classifiers for assessing the mental workload of the user across different days. It has been found that the performances of three different classifiers were significantly negatively impacted by the features time stability (Christensen et al. 2012). In this regards, the authors considered all the frequency bands (0.1–100 (Hz)) in the classifier's calibration stage, and probably the classifiers could not be able to select the features strictly related to the mental workload because of the big amount of information. Reliability of the system has great importance for a practical usability of such approach in real working contexts. In fact, the necessity to re-calibrate the system every day makes such kind of approach unusable in operational environments. An attempt to address such a kind of issues was investigated by Aricò and colleagues (Aricò et al. 2016), who proposed a new classification approach able to automatically optimize the features selection process, in order to make the classification performances of the considered mental state (e.g., mental workload) stable over time. In particular, the proposed new implementation of the *Stepwise Linear Discriminant Analysis* (SWLDA), that is the *automatic-stop SWLDA* (asSWLDA), is an extension of the *Linear Discriminant Analysis* (LDA, Fisher 1936), that has been proved to be one of the best outperforming linear classifiers (Craven et al. 2006; Krusienski et al. 2006; Rabbi et al. 2009). In fact, with respect to other linear methods, it has the advantage of having automatic features extraction in order to statistically remove the insignificant terms from the model. Particularly, the SWLDA regression consists in the combination of *forward* and *backward* stepwise analyses, where the input features are weighted by using ordinary least-squares regression to predict the target class label. The method starts by creating an initial model of the discriminant function, where the most statistically significant feature is added to the model for predicting the target labels ($pval_{ij} < \alpha_{ENTER}$), where $pval_{ij}$ represents the *p*-value of the *i*th feature at the *j*th iteration (in this case the first iteration). Then, at every new iteration, a new term is added to the model (if $pval_{ij} < \alpha_{ENTER}$). If there are not more features that satisfy this condition, a *backward* elimination analysis is performed to remove the least statistically significant feature (if $pval_{ij} > \alpha_{REMOVE}$). The standard implementation of the SWLDA algorithm uses $\alpha_{ENTER} = 0.05$ and $\alpha_{REMOVE} = 0.1$, and no constrains on the $\text{Iteraction}_{MAX}$ (predefined number of iterations) parameter are imposed, that is, the feature selection keeps going unless there are no more features satisfying the entry ($\alpha_{ENTER}$) and the removal ($\alpha_{REMOVE}$) conditions (Draper 1998b). Normally, it is possible to optimize a SWLDA regression by tuning all or some of the three parameters available in the algorithm ($\alpha_{ENTER}$, $\alpha_{REMOVE}$, and $\text{Iteraction}_{MAX}$). There are not standard procedures to choose these parameters, and they could be manually gauge. Despite the strength of the method, it is easy to realize how it would be difficult to set properly the parameters in order to optimize the algorithm.

In fact, since the significance associated to each feature is strictly related to the actual iteration and to the other features, this significance (pvalij) changes iteration by iteration, and it would result very difficult to impose a condition by manually trimming $\alpha_{ENTER}$ and $\alpha_{REMOVE}$. In addition, even if no constrains on the $\alpha_{ENTER}$ and the $\alpha_{REMOVE}$ parameters are imposed, the features are included in the model in order of significance (i.e., the first feature is the most significant one, and so on). As consequence, the value of the $\text{Iteration}_{MAX}$ parameter will affect the reliability of the classifier over time (optimum classifier, *underfitting* or *overfitting*). The expected result is that the more general the classification calibration would be, the higher the reliability of the algorithm over time will be (Vapnik 2000). For example, if the values of the parameters are too selective, that is $\alpha_{ENTER}$ and/or $\alpha_{REMOVE}$, and/or $\text{Iteration}_{MAX}$ values very low, the features added to the model will not be sufficient for predicting the target labels [*underfitting*, (von Luxburg and Schoelkopf 2008)]. On the contrary, if $\alpha_{ENTER}$ and/or $\alpha_{REMOVE}$, and/or $\text{Iteration}_{MAX}$ values are too high, most of the features added to the final model will be related to spurious differences between the classes of the training dataset, that are obviously not generalizable, so that, the reliability of the algorithm will decrease over time [*overfitting*, Vapnik (2000)]. The optimum solution to these problems would be to find out a criterion able to automatically stop the algorithm when the best number of features, $\#\text{Features}_{OPTIMUM}$, is added to the model.

The idea to fix such issues, for the proposed applications, was to define a criterion such that the number of the SWLDA iterations ($\text{Iteration}_{MAX}$) is: $\#\text{Features}_{UNDERFITTING} < \#\text{Features}_{OPTIMUM} < \#\text{Features}_{OVERFITTING}$. In other words, I wanted to make the SWLDA algorithm able to automatically select the right number of features ($\#\text{Features}_{OPTIMUM}$) to avoid both the *underfitting* and *overfitting* issues. In order to achieve such goal, I analyzed the *p-convergence*

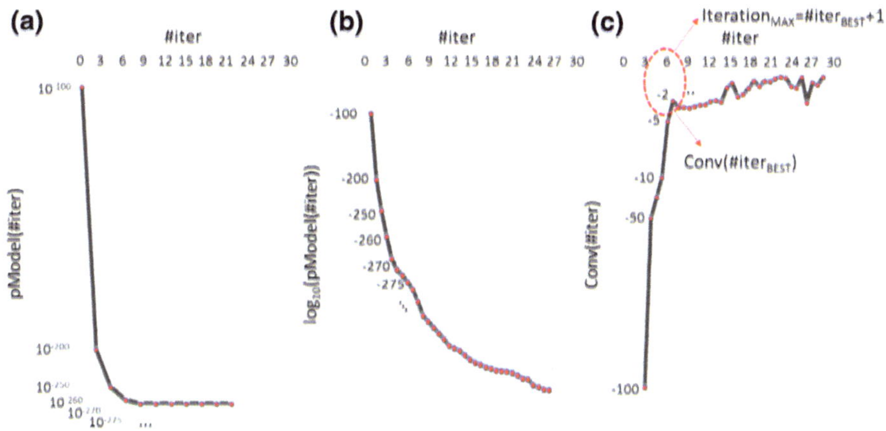


**Fig. 6.2.** Representation of the (**a**) *p*Model *vector*, the (**b**) $\log_{10}$ of the *p*Model vector and the (c) Conv function for each iteration, for a representative participant. In particular, in the figure (**c**) there are also showed (i) the Conv(#$\text{iter}_{\text{BEST}}$), in other words the lower distance of the Conv (#iter) function from the point (0,0) and (ii) the correspondent $\text{Iteration}_{\text{MAX}}$, that is #$\text{iter}_{\text{BEST}}$

function of the SWLDA classification model (Conv(#iter)). Since more the number of iterations increases (more features are added to the model), more the *p*-value of the model (*p*Model) decreases (tending to zero) with a decreasing exponential shape (convergence of the model), the idea of the best trade-off between the lowest number of features and the convergence of the model was to automatically stop the SWLDA (asSWLDA) algorithm in correspondence of the minimum distance from the (0, 0)-point, that is, when the Conv(#iter) function assumed the lowest distance from the origin. In Fig. 6.2a is reported the trend of the Conv(#iter) function for a representative participant.

The *p*Model values were returned by the MATLAB function of the SWLDA, and they have been collected for all the SWLDA iterations (*p*Model(#iter)). Then, the $\log_{10}$ of the *p*Model vector (log10(*p*Model(#iter)) has been applied to gather information about the *p*Model order. Figure 6.2b shows how the order of the model decreases with the number of interactions. The *p*Model parameter is available in the output of the standard SWLDA implementation and it provides information about the global significance of the model at the *j*th iteration.

To estimate the *p*-convergence function (Conv(#iter)) and the information about the size of *p*Model order, the $\log_{10}$ of the first-order differences between adjacent *p*Model elements has been calculated by the formula (6.7.1.1).

$$\text{Conv}(\#\text{iter}) = \log_{10}(p\text{Model}(\#\text{iter}+1)) - \log_{10}(p\text{Model}(\#\text{iter})) \quad (6.7.1.1)$$

Finally, the best $\text{Iteration}_{\text{MAX}}$ value has been identified as the number of iterations at which the Conv(#iter) assumed the lowest distance from the origin ($Iteration_{BEST}$), plus one because of the first-order differences:

$$\|\text{Conv}(\#\text{iter}_{\text{BEST}})\| = \min\|\text{Conv}(\#\text{iter})\| \quad (6.7.1.2)$$

$$\text{Iteration}_{\text{MAX}} = \#\text{iter}_{\text{BEST}} + 1 \quad (6..7.1.3)$$

In fact, the best condition for the aim of the considered studies (training assessment and workload evaluation over time) is to have a low numbers of features and, at the same time, the convergence of the model:

$$\log_{10}(pModel(\#\text{iter} + 1)) - \log_{10}(pModel(\#\text{iter})) = 0 \quad (6.7.1.4)$$

Figure 6.2c shows the graphical meaning of the $Iteration_{BEST}$ selection.

#### 6.7.6.1 asSWLDA Over-Time Stability

In order to assess the feasibility and reliability over-time (up to a month) of the asSWLDA algorithm, the performance of the standard SWLDA and the asSWLDA have been compared in terms of workload levels discrimination *within* the considered sessions, and *between* sessions spread out over time. The methods have been tested on users performing the MATB (see paragraph 3.1) under three different difficulty levels (Easy, Medium, and Hard). The MATB has been chosen due to its flexibility in modulating workload demand in a very controlled and standardized way (Aricò et al. 2015b). Finally, the algorithms have been tested in a simulated operational ATM environment (Aricò et al. 2016).

Figure 6.3 represents the mean AUC values ($z$-score normalized) of the two classifiers (standard SWLDA, blue line, and asSWLDA, red line) and the three cross-validation types (*Intra*, *Short term*, and *Medium term*).

In the *Intra* cross-validation, the calibration and testing dataset belonged to the same day; in the *Short term* cross-validation, the calibration triplets belonged to *Day 1* (*Day 2*), while the testing triplets to *Day 2* (*Day 1*); and finally, in the *Medium term* cross-validation, the calibration triplets belonged to *Day 1* or *Day 2* (*Day N*), and the testing triplets to *Day N* (*Day 1* or *Day 2*). $N$ is equal to 7 or 30 when the stability was evaluated over a week or over a month, respectively. The ANOVAs related to the comparison of the two classifiers along the three cross-validation types, highlighted no significant differences in the Intra condition ([asSWLDA vs standard SWLDA]: $F(1, 18) = 0.32$; $p = 0.57$). On the contrary, significant differences in the Short-term and the Medium-term conditions have been found (Short term [asSWLDA vs. standard SWLDA]: $F(1, 18) = 4.22$; $p = 0.05$; Medium term [asSWLDA vs. standard SWLDA]: $F(1, 18) = 6.7$; $p = 0.02$). The ANOVAs on the AUC values of the three cross-validation types for each classifier reported the following results. No significant differences between the cross-validation types have been highlighted for the asSWLDA classifier (asSWLDA [Intra, Short term, Medium term]: $F(2, 27) = 0.62$; $p = 0.54$). On the contrary, for the standard SWLDA a significant effect between the cross-validation types has been highlighted (standard SWLDA [Intra, Short term, Medium term]:

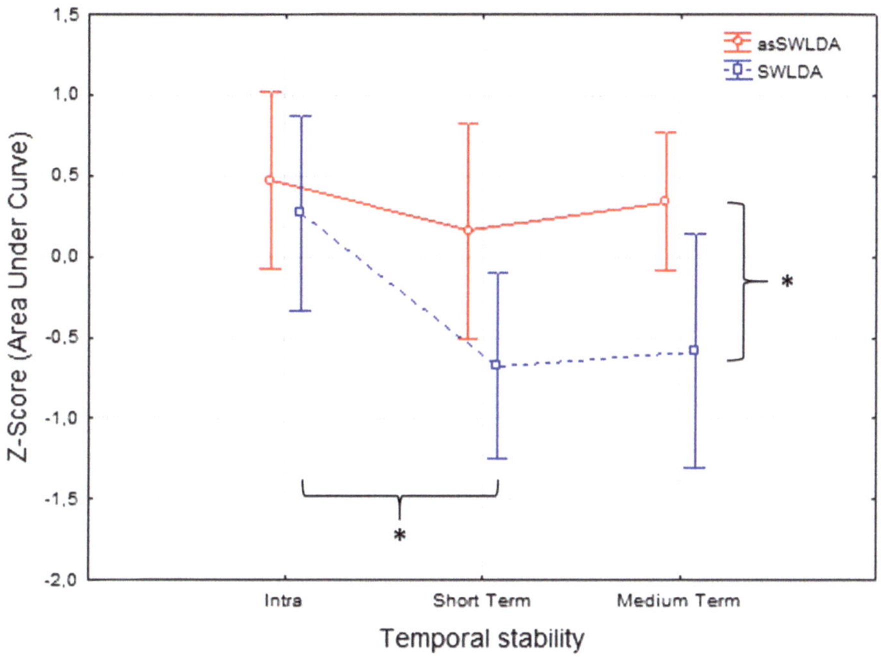


**Fig. 6.3.** Error bars (CI = 0.95) related to the AUC values (*z*-score corrected) of the classifier by using the two classifiers (asSWLDA—*red line*, standard SWLDA—*blue line*) and the three cross-validation types (Intra, Short term, Medium term)

$F(2, 27) = 3.57$; $p = 0.04$). In particular, the post hoc test highlighted a significant decrement of the AUC values between the Intra and the Short term and the Medium term cross-validations (all $p < 0.05$). These results demonstrated that the asSWLDA classifier allowed to maintain the performance of the measure stable across days (no significant difference between the cross-validation types). On the contrary, the standard SWLDA classifier suffered of a significant decrement of the performances across days. In fact, since the second day (short term) the performance of the asSWLDA (red line) outperformed the standard SWLDA (blue line).

#### 6.7.6.2 Mental Workload Discrimination

As consequence of the results described above, we evaluated how well the asSWLDA was able to discriminate the three MATB conditions (Easy, Medium, and Hard) across different days. Figure 6.4 shows the time representation of the $W_{EEG}$ index for a representative participant along each difficulty level (Easy, Medium, and Hard).

ANOVA analyses showed that the $W_{EEG}$ values related to the three difficulty levels (Easy, Medium, and Hard) were significantly different for each cross-validation type

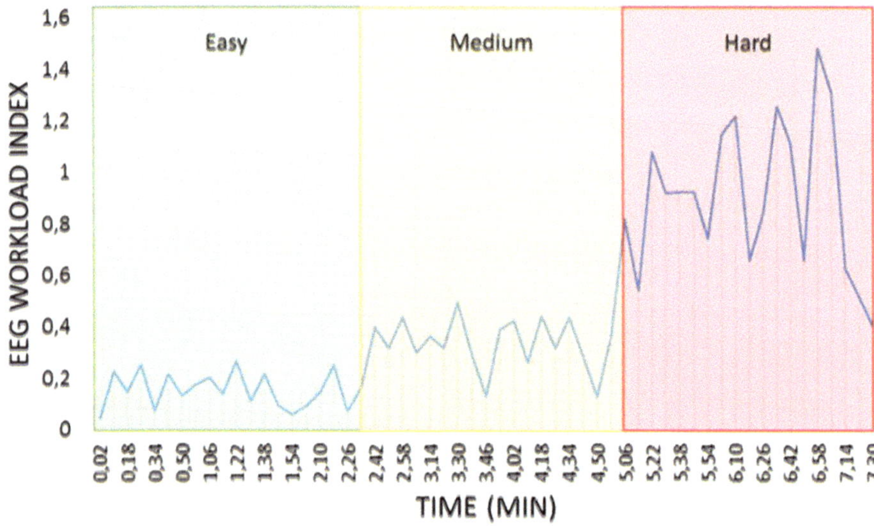


**Fig. 6.4.** Time representation [time resolution of 8 (s)] of the $W_{EEG}$ index evaluated for the three task conditions. From *left* to *right*: Easy, Medium, and Hard condition for a representative participant

(Intra: $F(2, 27) = 417.65$; $p = 10^{-6}$; Short term: $F(2, 27) = 360.9$; $p = 10^{-6}$; Medium term: $F(2, 27) = 459.78$; $p = 10^{-6}$). The post hoc test confirmed that the $W_{EEG}$ distributions related to the three difficulty levels have been significantly different from each other at every cross-validation type (all $p < 10^{-4}$).

#### 6.7.6.3 Discussion and Conclusions

In this study has been demonstrated that (1) the asSWLDA significantly outperformed the standard implementation of the SWLDA; (2) the asSWLDA algorithm's performance remained high across different days (up to a month) without any recalibrations, so it is reliable over time; (3) the algorithm was able to significantly differentiate three mental workload levels. Therefore, the asSWLDA has been used for the studies presented within the book.

## 6.8 Self-reported Questionnaires

### *6.8.1 NASA-Task Load Index (NASA-TLX)*

The NASA–*Task Load Index* (NASA–TLX) is a widely used, subjective, multidimensional assessment tool that rates perceived workload in order to assess a task, system, team's effectiveness or other performance aspects. It was developed in the 80' by the Human Performance Group at NASA's Ames Research Centre over a three-year development cycle that included more than 40 laboratory simulations (Hart and Staveland 1988). It has been used in a variety of domains, including aviation, health care, and other complex socio-technical domains. The total workload score is calculated by the combination of six factors:

- *Mental Demand*: how much mental and perceptual activity was required (e.g., thinking, deciding, calculating, remembering, looking, searching, etc.)? Was the task easy or demanding, simple or complex, exacting or forgiving?
- *Physical Demand*: how much physical activity was required (e.g., pushing, pulling, turning, controlling, activating, etc.)? Was the task easy or demanding, slow or brisk, slack or strenuous, restful or laborious?
- *Temporal Demand*: how much time pressure did the subject feel due to the rate or pace at which the tasks or task's elements occurred? Was the pace slow and leisurely or rapid and frantic?
- *Performance*: how successful did the subject think he/she was in accomplishing the goals of the task set by the experimenter (or him/herself)? How satisfied was the subject with his/her performance in accomplishing these goals?
- *Effort*: how hard did the subject have to work (mentally and physically) to accomplish his/her level of performance?
- *Frustration*: how insecure, discouraged, irritated, stressed, and annoyed versus secure, gratified, content, relaxed, and complacent did the subject feel during the task?

At the end of the task or during the execution, the NASA-TLX is presented and the participants have to rate each factor (from 0 to 100).

Later, the factors are compared in pairs to each others, and the number of times each factor is chosen is the weight by which the previous rates will be multiplied. The linear combination of the weighted rates is then divided by 15 (number of total comparisons) and thus the NASA-TLX total score (value from 0 to 100) will be provided. The NASA-TLX can be administered using paper sheets or digital interfaces. An example of interface is reported in Fig. 6.5. While there are multiple ways to administer the NASA-TLX, some may change the results of the test. One study showed that the paper sheets version led to less cognitive workload than the digital one (Noyes and Bruneau 2007). To overcome the delay in administrating the test, the digital version can be used to submit the test immediately after the task is completed in order to capture accurate workload perception.

### 6.8.2 Instantaneous Self Assessment (ISA)

*Instantaneous Self Assessment* (ISA) is a technique that has been developed as a measure of workload to provide immediate subjective ratings of work demands during the performance of primary work tasks (Tattersall and Foord 1996). The ISA

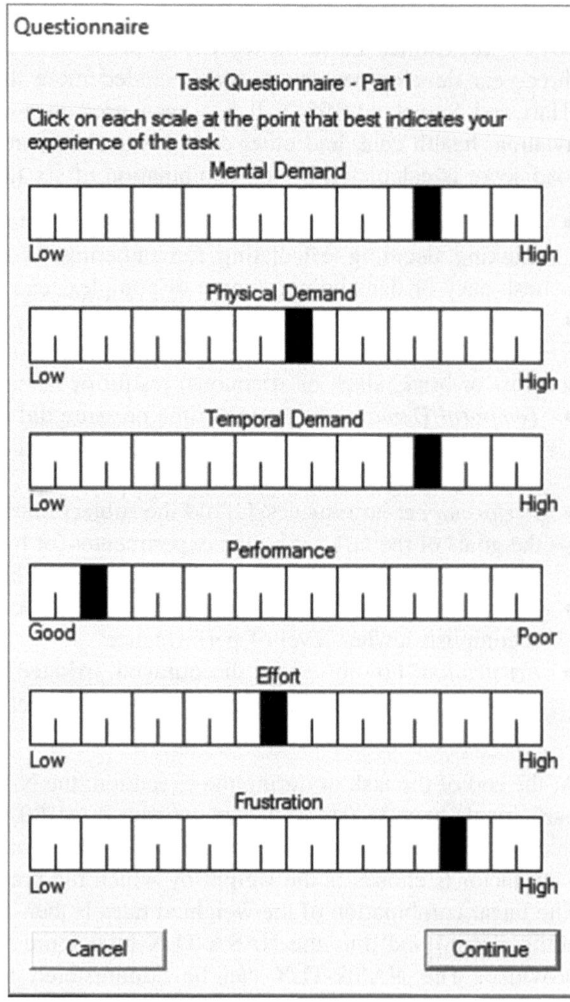


**Fig. 6.5.** Example of NASA—TLX interface. The total workload index is a weighted combination of six factors: mental demand, physical demand, temporal demand, performance, effort, and frustration

was originally developed by the *National Air Traffic Service* (NATS) in the UK for use in the assessment of air traffic controller's mental workload during the design of future ATM systems (Kirwan 1998; Kirwan et al. 2005). It has then been used by other UK organizations, including the *Air Traffic Control Evaluation Unit* (ATCEU) at Bournemouth airport, the *Defence Evaluation and Research Agency* (DERA) in Portsdown and the *Royal Navy's research organisation* (now part of *QinetiQ*), and also European organizations, the *Eurocontrol Experimental Centre* (EEC), the *Centre d'études de la navigation aérienne* (CENA) and by the *Rijksluchtvaartdienst* (RLD) of Netherland (now *Civil Aviation Authority*—CAA). ISA involves participants that self-rate their workload during a task (normally every two minutes) on a scale of 1 (low) to 5 (high). Kirwan et al. (2005) used the following (Fig. 6.6) ISA definitions to assess air traffic controllers' (ATC) workload.

| Level | Workload Heading | Spare Capacity | Description |
|---|---|---|---|
| 5 | Excessive | None | Behind on tasks; losing track of the full picture |
| 4 | High | Very Little | Non-essential tasks suffering. Could not work at this level very long. |
| 3 | Comfortable Busy Pace | Some | All tasks well in hand. Busy but stimulating pace. Could keep going continuously at this level. |
| 2 | Relaxed | Ample | More than enough time for all tasks. Active on ATC task less than 50% of the time. |
| 1 | Under-Utilised | Very Much | Nothing to do. Rather boring. |

**Fig. 6.6.** Example of ISA scale. When the ISA scale is presented to the user, he\she has to select a value from 1 (very low) to 5 (very high) in order to score the perception of the workload on that moment

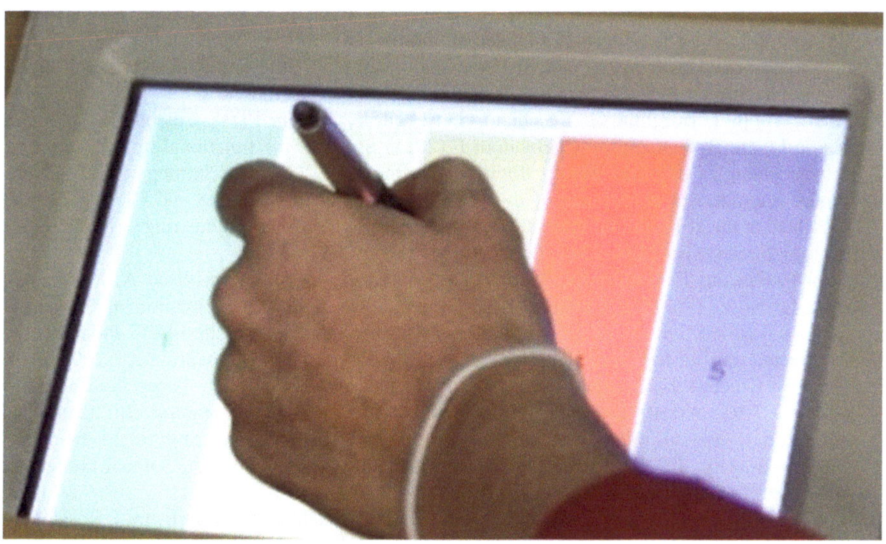

**Fig. 6.7.** The ISA test is typically presented to the participants in the form of a color-coded keypad. The keypad flashes and sounds when the workload rating is required, and the participant simply pushes the button that corresponds to their perception of workload

Typically, the ISA scale is presented to the participants in the form of a color-coded keypad. The keypad flashes and sounds when the workload rating is required, and the participant simply push the button that corresponds to their workload perception (Fig. 6.7). Alternatively, the workload ratings can be requested and acquired verbally. The appeal of the ISA technique lies in its low resource usage and its low intrusiveness. In fact, the output of the ISA (values from 1 to 5) allows a workload profile for the task under analysis to be constructed and it is very quick in its application as data collection occurs during the trial. While the technique might be a bit obtrusive to the primary task, it is probably the least intrusive of the online workload assessment techniques.

## References

Aricò P, Borghini G, Di Flumeri G, Colosimo A, Graziani I, Imbert JP, Granger G, Benhacene R, Terenzi M, Pozzi S, Babiloni F (2015) Reliability over time of EEG-based mental workload evaluation during Air Traffic Management (ATM) tasks. In: Conference proceedings: annual international conference of the IEEE Engineering in Medicine and Biology Society. IEEE Engineering in Medicine and Biology Society. Annual conference 2015, pp 7242–7245. doi:10.1109/EMBC.2015.7320063

Aricò P, Borghini G, Di Flumeri G, Colosimo A, Pozzi S, Babiloni F (2016). A passive brain-computer interface application for the mental workload assessment on professional air traffic controllers during realistic air traffic control tasks. Prog Brain Res 228:295–328. doi:10.1016/bs.pbr.2016.04.021

Astolfi L, de Vico Fallani F, Cincotti F, Mattia D, Marciani MG, Bufalari S, Salinari S, Colosimo A, Ding L, Edgar JC, Heller W, Miller GA, He B, Babiloni F (2007) Imaging functional brain connectivity patterns from high-resolution EEG and fMRI via graph theory. Psychophysiology 44:880–893. doi:10.1111/j.1469-8986.2007.00556.x

Astolfi L, Toppi J, Borghini G, Vecchiato G, Isabella R, Fallani FDV, Cincotti F, Salinari S, Mattia D, He B, Caltagirone C, Babiloni F (2011) Study of the functional hyperconnectivity between couples of pilots during flight simulation: an EEG hyperscanning study. In: 2011 annual international conference of the IEEE Engineering in Medicine and Biology Society. Presented at the 2011 annual international conference of the IEEE Engineering in Medicine and Biology Society, pp 2338–2341. doi:10.1109/IEMBS.2011.6090654

Babiloni F, Cincotti F, Babiloni C, Carducci F, Mattia D, Astolfi L, Basilisco A, Rossini PM, Ding L, Ni Y, Cheng J, Christine K, Sweeney J, He B (2005) Estimation of the cortical functional connectivity with the multimodal integration of high-resolution EEG and fMRI data by directed transfer function. NeuroImage 24:118–131. doi:10.1016/j.neuroimage.2004.09.036

Babiloni F, Cincotti F, Mattia D, Mattiocco M, De Vico Fallani F, Tocci A, Bianchi L, Marciani MG, Astolfi L (2006) Hypermethods for EEG hyperscanning. In: Conference proceedings: annual international conference of the IEEE Engineering in Medicine and Biology Society. IEEE Engineering in Medicine and Biology Society. Annual conference, vol 1, pp 3666–3669. doi:10.1109/IEMBS.2006.260754

Babiloni C, Buffo P, Vecchio F, Marzano N, Del Percio C, Spada D, Rossi S, Bruni I, Rossini PM, Perani D (2012) Brains "in concert": frontal oscillatory alpha rhythms and empathy in professional musicians. NeuroImage 60:105–116. doi:10.1016/j.neuroimage.2011.12.008

Baccalá LA, Sameshima K (2001) Partial directed coherence: a new concept in neural structure determination. Biol Cybern 84:463–474. doi:10.1007/PL00007990

Bamber D (1975) The area above the ordinal dominance graph and the area below the receiver operating characteristic graph. J Math Psychol 12:387–415. doi:10.1016/0022-2496(75)90001-2

Berka C, Levendowski DJ, Lumicao MN, Yau A, Davis G, Zivkovic VT, Olmstead RE, Tremoulet PD, Craven PL (2007) EEG correlates of task engagement and mental workload in vigilance, learning, and memory tasks. Aviat Space Environ Med 78:B231–B244

Bezerianos A, Sun Y, Chen Y, Woong KF, Taya F, Arico P, Borghini G, Babiloni F, Thakor N (2015) Cooperation driven coherence: brains working hard together. In: 2015 37th annual international conference of the IEEE Engineering in Medicine and Biology Society (EMBC). Presented at the 2015 37th annual international conference of the IEEE Engineering in Medicine and Biology Society (EMBC), pp 4696–4699. doi:10.1109/EMBC.2015.7319442

Blankertz B, Tangermann M, Vidaurre C, Fazli S, Sannelli C, Haufe S, Maeder C, Ramsey L, Sturm I, Curio G, Müller K-R (2010) The Berlin brain–computer interface: non-medical uses of BCI technology. Front Neurosci 4. doi:10.3389/fnins.2010.00198

Brodmann's (2006) Localisation in the Cerebral Cortex

Burgess A, Gruzelier J (1993) Individual reliability of amplitude distribution in topographical mapping of EEG. Electroencephalogr Clin Neurophysiol 86:219–223

Burgess AP (2013a) On the interpretation of synchronization in EEG hyperscanning studies: a cautionary note. Front Hum Neurosci 7. doi:10.3389/fnhum.2013.00881

Cartocci G, Maglione AG, Vecchiato G, Di Flumeri G, Colosimo A, Scorpecci A, Marsella P, Giannantonio S, Malerba P, Borghini G, Arico P, Babiloni F (2015) Mental workload estimations in unilateral deafened children. In: Conference proceedings: annual international conference of the IEEE Engineering in Medicine and Biology Society. IEEE Engineering in Medicine and Biology Society. Annual conference, pp 1654–1657. doi:10.1109/EMBC.2015.7318693

Chandrasekaran V, Jordan MI (2013) Computational and statistical tradeoffs via convex relaxation. Proc Natl Acad Sci 110:E1181–E1190. doi:10.1073/pnas.1302293110

Cherkas A, Abrahamovych O, Golota S, Nersesyan A, Pichler C, Serhiyenko V, Knasmüller S, Zarkovic N, Eckl P (2015) The correlations of glycated hemoglobin and carbohydrate metabolism parameters with heart rate variability in apparently healthy sedentary young male subjects. Redox Biol 5:301–307. doi:10.1016/j.redox.2015.05.007

Christensen JC, Estepp JR, Wilson GF, Russell CA (2012) The effects of day-to-day variability of physiological data on operator functional state classification. NeuroImage 59:57–63. doi:10.1016/j.neuroimage.2011.07.091

Craven PL, Belov N, Tremoulet P, Thomas M, Berka C, Levendowski D, Davis G (2006) Cognitive workload gauge development: comparison of real-time classification methods. Foundations of Augmented Cognition, 75–84

Daly JJ, Wolpaw JR (2008) Brain-computer interfaces in neurological rehabilitation. Lancet Neurol 7:1032–1043. doi:10.1016/S1474-4422(08)70223-0

De Vico Fallani F, Nicosia V, Sinatra R, Astolfi L, Cincotti F, Mattia D, Wilke C, Doud A, Latora V, He B, Babiloni F (2010) Defecting or not defecting: how to "read" human behavior during cooperative games by EEG measurements. PloS One 5:e14187. doi:10.1371/journal.pone.0014187

Decatur S, Goldreich O, Ron D (2000) Computational sample complexity. SIAM J Comput 29:854–879. doi:10.1137/S0097539797325648

Delorme A, Makeig S (2004) EEGLAB: an open source toolbox for analysis of single-trial EEG dynamics including independent component analysis. J Neurosci Methods 134:9–21. doi:10.1016/j.jneumeth.2003.10.009

Di Flumeri G, Aricò P, Borghini G, Colosimo A, Babiloni F (2016a) A new regression-based method for the eye blinks artifacts correction in the EEG signal, without using any EOG channel. In: 2016 38th Annual International Conference of the IEEE Engineering in Medicine and Biology Society (EMBC). Presented at the 2016 38th Annual International Conference of the IEEE Engineering in Medicine and Biology Society (EMBC), pp. 3187–3190. doi: 10.1109/EMBC.2016.7591406

Di Flumeri G, Herrero MT, Trettel A, Cherubino P, Maglione AG, Colosimo A, Moneta E, Peparaio M, Babiloni F (2016b) EEG frontal asymmetry related to pleasantness of olfactory

stimuli in young subjects. In: Nermend K, Łatuszyńska M (eds) Selected Issues in experimental economics. Springer International Publishing, Cham, pp 373–381
Ding L, Lai Y, He B (2005) Low resolution brain electromagnetic tomography in a realistic geometry head model: a simulation study. Phys Med Biol 50:45–56
Draper NR (1998a) Applied regression analysis bibliography update 1994–97. Commun Stat Theory Methods 27:2581–2623. doi:10.1080/03610929808832244
Draper NR (1998b) Applied regression analysis. Commun Stat Theory Methods 27:2581–2623. doi:10.1080/03610929808832244
Dumas G, Nadel J, Soussignan R, Martinerie J, Garnero L (2010) Inter-brain synchronization during Social Interaction. PLoS ONE 5:e12166. doi:10.1371/journal.pone.0012166
Dumas G, Chavez M, Nadel J, Martinerie J (2012) Anatomical connectivity influences both intra- and inter-brain synchronizations. PLoS ONE 7:e36414. doi:10.1371/journal.pone.0036414
Elul R (1969) Gaussian behavior of the electroencephalogram: changes during performance of mental task. Science 164:328–331
Faes L, Nollo G (2011) Multivariate frequency domain analysis of causal interactions in physiological time series. INTECH Open Access Publisher, Croatia
Fallani FDV, Toppi J, Lanzo CD, Vecchiato G, Astolfi L, Borghini G, Mattia D, Cincotti F, Babiloni F (2012) Redundancy in functional brain connectivity from eeg recordings. Int J Bifurc Chaos 22:1250158-1:1250158-8. doi:10.1142/S0218127412501581
Fisher RA (1936) The use of multiple measurements in taxonomic problems. Ann Eugen 7: 179–188. doi:10.1111/j.1469-1809.1936.tb02137.x
French CC, Beaumont JG (1984) A critical review of EEG coherence studies of hemisphere function. Int J Psychophysiol Off J Int Organ Psychophysiol 1:241–254
Fukunaga K (1990) Statistical pattern recognition, second, 2nd edn. Academic Press, INC, San Diego
Garrett D, Peterson DA, Anderson CW, Thaut MH (2003) Comparison of linear, nonlinear, and feature selection methods for EEG signal classification. IEEE Trans Neural Syst Rehabil Eng 11:141–144. doi:10.1109/TNSRE.2003.814441
Giraudet L, Imbert J-P, Bérenger M, Tremblay S, Causse M (2015) The neuroergonomic evaluation of human machine interface design in air traffic control using behavioral and EGG/ERP measures. Behav Brain Res 294:246–253. doi:10.1016/j.bbr.2015.07.041
Gordan R, Gwathmey JK, Xie L-H (2015) Autonomic and endocrine control of cardiovascular function. World J Cardiol 7:204–214. doi:10.4330/wjc.v7.i4.204
Granger CWJ (2004) Time series analysis, cointegration, and applications. Am Econ Rev 94:421–425. doi:10.1257/0002828041464669
Gratton G, Coles MG, Donchin E (1983) A new method for off-line removal of ocular artifact. Electroencephalogr Clin Neurophysiol 55:468–484
de Peralta Grave, Menendez R, Andino SG (1999) Distributed source models: standard solutions and new developments. In: Uhl C (ed) Analysis of neurophysiological brain functioning. Springer, Berlin, pp 176–201
Hansen PC (1992) Analysis of discrete Ill-posed problems by means of the L-curve. SIAM Rev 34:561–580. doi:10.1137/1034115
Harris FJ (1978) On the use of windows for harmonic analysis with the discrete fourier transform. Proc IEEE 66:51–83. doi:10.1109/PROC.1978.10837
Hart SG, Staveland LE (1988) Development of NASA-TLX (Task Load Index): results of empirical and theoretical research. Human Mental Workload, North-Holland, pp 139–183
He B, Hori J, Babiloni F (2006) Electroencephalography (EEG): inverse problems. In: Wiley Encyclopedia of Biomedical Engineering. Wiley, London
Hinton GE, Salakhutdinov RR (2006) Reducing the dimensionality of data with neural networks. Science 313:504–507. doi:10.1126/science.1127647
Hoffmann U, Vesin J-M, Ebrahimi T, Diserens K (2008) An efficient P300-based brain–computer interface for disabled subjects. J Neurosci Methods, Brain-Computer Interfaces (BCIs) 167:115–125. doi:10.1016/j.jneumeth.2007.03.005

Jordan MI, Mitchell TM (2015) Machine learning: trends, perspectives, and prospects. Science 349:255–260. doi:10.1126/science.aaa8415

Kaber DB, Perry CM, Segall N, Sheik-Nainar MA (2007) Workload state classification with automation during simulated air traffic control. Int J Aviat Psychol 17:371–390. doi:10.1080/10508410701527860

Kawasaki M, Yamada Y, Ushiku Y, Miyauchi E, Yamaguchi Y (2013) Inter-brain synchronization during coordination of speech rhythm in human-to-human social interaction. Sci Rep 3:1692. doi:10.1038/srep01692

Kirwan B (1998) Human error identification techniques for risk assessment of high risk systems–Part 1: review and evaluation of techniques. Appl Ergon 29:157–177

Kirwan B, Rodgers M, Schäfer D (2005) Human factors impacts in air traffic management. Ashgate, Vermont, Burlington

Klimesch W (1999) EEG alpha and theta oscillations reflect cognitive and memory performance: a review and analysis. Brain Res Rev 29:169–195. doi:10.1016/S0165-0173(98)00056-3

Kong W, Lin W, Babiloni F, Hu S, Borghini G (2015) Investigating driver fatigue versus alertness using the granger causality network. Sensors 15:19181–19198. doi:10.3390/s150819181

Kong W, Zhou Z, Jiang B, Babiloni F, Borghini G (2017) Assessment of driving fatigue based on intra/inter-region phase synchronization. Neurocomputing 219:474–482. doi:10.1016/j.neucom.2016.09.057

Krusienski DJ, Sellers EW, Cabestaing F, Bayoudh S, McFarland DJ, Vaughan TM, Wolpaw JR (2006) A comparison of classification techniques for the P300 Speller. J Neural Eng 3:299–305. doi:10.1088/1741-2560/3/4/007

Lachaux JP, Rodriguez E, Martinerie J, Varela FJ (1999) Measuring phase synchrony in brain signals. Hum Brain Mapp 8:194–208

Lal SKL, Craig A (2002) Driver fatigue: electroencephalography and psychological assessment. Psychophysiology 39:313–321. doi:10.1017/S0048577201393095

Lemm S, Blankertz B, Dickhaus T, Müller K-R (2011) Introduction to machine learning for brain imaging. NeuroImage Multivar Decoding Brain Reading 56:387–399. doi:10.1016/j.neuroimage.2010.11.004

Lindenberger U, Li S-C, Gruber W, Müller V (2009) Brains swinging in concert: cortical phase synchronization while playing guitar. BMC Neurosci 10:22. doi:10.1186/1471-2202-10-22

Lotte F, Congedo M, Lécuyer A, Lamarche F, Arnaldi B (2007) A review of classification algorithms for EEG-based brain–computer interfaces. J Neural Eng 4:R1. doi:10.1088/1741-2560/4/2/R01

McEvoy LK, Smith ME, Gevins A (2000) Test-retest reliability of cognitive EEG. Clin Neurophysiol Off J Int Fed Clin Neurophysiol 111:457–463

McFarland DJ, Wolpaw JR (2005) Sensorimotor rhythm-based brain-computer interface (BCI): feature selection by regression improves performance. IEEE Trans Neural Syst Rehabil Eng 13:372–379. doi:10.1109/TNSRE.2005.848627

McFarland DJ, Krusienski DJ, Wolpaw JR (2006) Brain-computer interface signal processing at the Wadsworth center: mu and sensorimotor beta rhythms. Prog Brain Res 159:411–419. doi:10.1016/S0079-6123(06)59026-0

Mnih V, Badia AP, Mirza M, Graves A, Lillicrap TP, Harley T, Silver D, Kavukcuoglu K (2016) asynchronous methods for deep reinforcement learning. ArXiv160201783 Cs

Montague PR (1999) Reinforcement learning: an introduction. In: Sutton RS, Barto AG Trends Cogn Sci 3:360. doi:10.1016/S1364-6613(99)01331-5

Murphy B, Talukdar PP, Mitchell T (2012) Learning effective and interpretable semantic models using non-negative sparse embedding. International conference on computational linguistics (COLING), Mumbai, India

Naeem M, Prasad G, Watson DR, Kelso JAS (2012) Electrophysiological signatures of intentional social coordination in the 10–12 Hz range. NeuroImage 59:1795–1803. doi:10.1016/j.neuroimage.2011.08.010

Nicolas-Alonso LF, Gomez-Gil J (2012) Brain computer interfaces, a review. Sensors 12:1211–1279. doi:10.3390/s120201211

Noyes JM, Bruneau DPJ (2007) A self-analysis of the NASA-TLX workload measure. Ergonomics 50:514–519. doi:10.1080/00140130701235232
Nunez PL, Cutillo BA (1995) Neocortical dynamics and human EEG rhythms. Oxford University Press, New York
Parasuraman R, Rizzo M (2008) Neuroergonomics: the brain at work, 1 edn. Oxford University Press, New York
Rabbi AF, Ivanca K, Putnam AV, Musa A, Thaden CB, Fazel-Rezai R (2009) Human performance evaluation based on EEG signal analysis: a prospective review. Conference proceedings: annual international conference of the IEEE Engineering in Medicine and Biology Society. IEEE Engineering in Medicine and Biology Society. Annual conference, 1879–1882. doi:10.1109/IEMBS.2009.5333877
Rasmussen J (1986) Information processing and human-machine interaction: an approach to cognitive engineering. North-Holland
Romero S, Mañanas MA, Barbanoj MJ (2008) A comparative study of automatic techniques for ocular artifact reduction in spontaneous EEG signals based on clinical target variables: A simulation case. Comput Biol Med 38:348–360. doi:10.1016/j.compbiomed.2007.12.001
Sakkalis V (2011) Review of advanced techniques for the estimation of brain connectivity measured with EEG/MEG. Comput Biol Med 41:1110–1117. doi:10.1016/j.compbiomed.2011.06.020
Salinsky MC, Oken BS, Morehead L (1991) Test-retest reliability in EEG frequency analysis. Electroencephalogr Clin Neurophysiol 79:382–392
Sänger J, Müller V, Lindenberger U (2013) Directionality in hyperbrain networks discriminates between leaders and followers in guitar duets. Front Hum Neurosci 7:234. doi:10.3389/fnhum.2013.00234
Scerbo MW, Freeman FG, Mikulka PJ, Parasuraman R, Di Nocera F, Prinzel LJ III (2001) The efficacy of psychophysiological measures for implementing adaptive technology. NASA langley technical report server
Shalev-Shwartz S, Zhang T (2013) Stochastic dual coordinate ascent methods for regularized loss minimization. J Mach Learn Res 14:567–599
Steel RG, Torrie JH, Dickey DA (1996) Principles and procedures of statistics: a biometrical approach, 3 Sub edn. McGraw-Hill Companies, New York
Tass P, Rosenblum MG, Weule J, Kurths J, Pikovsky A, Volkmann J, Schnitzler A, Freund H-J (1998) Detection of n: m phase locking from noisy data: application to magnetoencephalography. Phys Rev Lett 81:3291–3294. doi:10.1103/PhysRevLett.81.3291
Tattersall AJ, Foord PS (1996) An experimental evaluation of instantaneous self-assessment as a measure of workload. Ergonomics 39:740–748. doi:10.1080/00140139608964495
Taya F, Sun Y, Babiloni F, Thakor N, Bezerianos A (2015) Brain enhancement through cognitive training: a new insight from brain connectome. Front. Syst. Neurosci. 9:44. doi:10.3389/fnsys.2015.00044
Tognoli E (2008) EEG coordination dynamics: neuromarkers of social coordination. In: Fuchs A, Jirsa VK (eds) Coordination: neural, behavioral and social dynamics, understanding complex systems. Springer, Berlin, Heidelberg, pp 309–323. doi:10.1007/978-3-540-74479-5_15
Townsend G, LaPallo BK, Boulay CB, Krusienski DJ, Frye GE, Hauser CK, Schwartz NE, Vaughan TM, Wolpaw JR, Sellers EW (2010) A novel P300-based brain–computer interface stimulus presentation paradigm: moving beyond rows and columns. Clin Neurophysiol 121:1109–1120. doi:10.1016/j.clinph.2010.01.030
Vapnik VN (2000) The nature of statistical learning theory. Springer, New York
Vidaurre C, Krämer N, Blankertz B, Schlögl A (2009) Time domain parameters as a feature for EEG-based brain–computer interfaces. Neural Netw Brain-Machine Interface 22:1313–1319. doi:10.1016/j.neunet.2009.07.020
von Luxburg U, Schoelkopf B (2008) Statistical learning theory: models, concepts, and results. ArXiv08104752 Math Stat
White DW, Raven PB (2014) Autonomic neural control of heart rate during dynamic exercise: revisited. J Physiol 592:2491–2500. doi:10.1113/jphysiol.2014.271858

Wilson GF (2002) An analysis of mental workload in pilots during flight using multiple psycho-physiological measures. Int J Aviat Psychol 3–18

Wilson GF, Russell CA (2003) Real-time assessment of mental workload using psychophysiological measures and artificial neural networks. Hum Factors J Hum Factors Ergon Soc 45:635–644. doi:10.1518/hfes.45.4.635.27088

Zander TO, Kothe C (2011) Towards passive brain–computer interfaces: applying brain–computer interface technology to human–machine systems in general. J Neural Eng 8:025005. doi:10.1088/1741-2560/8/2/025005

# Chapter 7
# Applications

In this chapter, some industrial applications are described. In particular, for the investigation of the different mental states (mental workload, cognitive control behavior, and training), the triangulation approach has been used (Bekhet and Zauszniewski 2012; Jick 1979). In other words, information coming from different sources, such as behavioral data, self-reported questionnaires, neurophysiological signals, were combined in order to better analyze and evaluate the considered cognitive phenomena. Multimodal approaches are very important for avoiding wrong interpretation of the results. For example, the users could achieve high performance in terms of task execution, but show high mental demand from a cognitive perspective when dealing with the task itself. For each mental status, preliminary experiments were performed in controlled settings (i.e., Lab), and then the second phase of the experiments has been run in more ecological contexts. Thus, the purpose of the following studies was to investigate the effectiveness of the developed methodologies, and to define specific neurometrics for the analysis of the considered mental states.

## 7.1 Mental Workload Evaluation

### *7.1.1 Air Traffic Controllers' Mental Workload Evaluation*

The aim of this study was to test if the asSWLDA algorithm, the neurometric, and the methodology developed in the previous studies for the workload evaluation were able to: (1) evaluate the workload levels in real ATM settings, (2) highlight the differences in terms of workload between groups of ATCOs. In order to compare the groups, a workload index has been defined, as the ratio between the frontal theta and the parietal alpha PSDs (*ThetaF/AlphaP Index*). Furthermore, the asSWLDA



G. Borghini et al., *Industrial Neuroscience in Aviation*,
Biosystems & Biorobotics 18, DOI 10.1007/978-3-319-58598-7_7

algorithm has been used to compute an *EEG-based Mental Workload Index* ($W_{EEG}$) able to track the workload during the execution of the ATC task.

*Participants*

Thirty-seven professional ATCOs from the *École Nationale de l'Aviation Civile* (ENAC) of Toulouse (France) have been involved in this study. They have been selected in order to have homogeneous experimental groups in terms of age and expertise. In particular, two groups of controllers have been defined: ATC Experts (40.41 ± 5.54), and ATC Students (23 ± 1.95). ATCOs have been asked to take part in a simulated ATC simulation where the same ATM scenario has been proposed.

*Results*

The ANOVAs performed on the performance and workload indexes (*between* factor RANK; 2 levels: experts and students) reported significant differences between the two groups. In particular, the ATC Experts reacted and provided information to the pilots significantly faster [F (1, 31) = 4.7621; p = 0.03679], and experienced a significant lower [F(1, 35) = 8.8145; p = 0.00536] workload than the ATC Students (Figs. 7.1, 7.2, respectively).

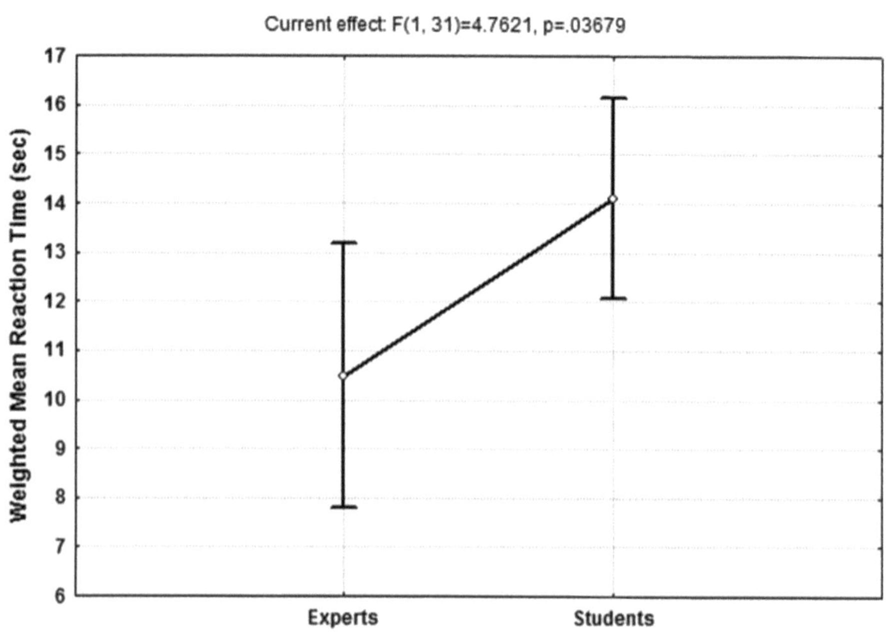


**Fig. 7.1** Differences between the ATC experts and ATC students in terms of Weighted Mean Reaction Time (WMRT). The ANOVA analysis showed a significant difference (p = 0.037) between the two groups

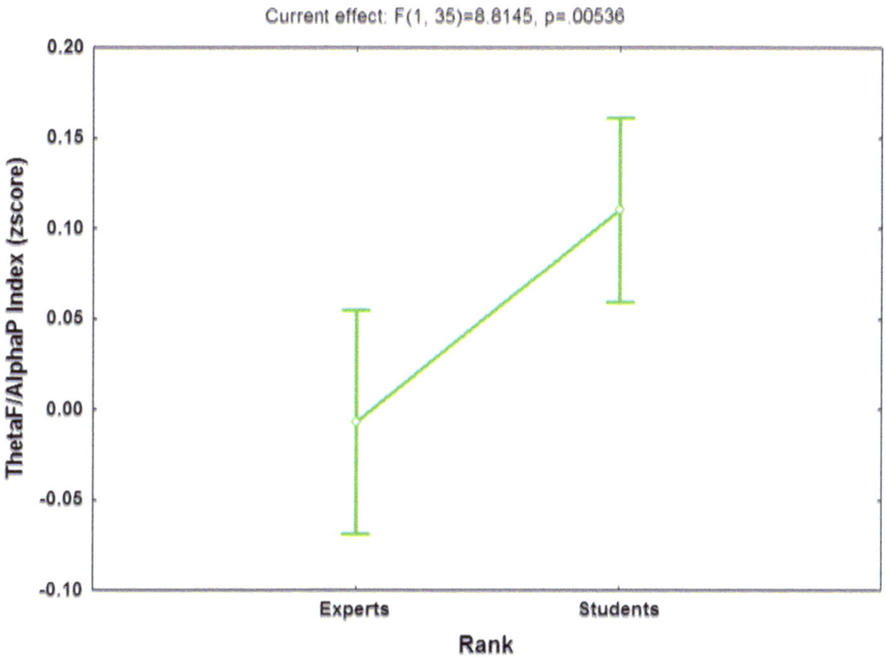


**Fig. 7.2** The ANOVA on the *ThetaF/AlphaP Index* reported a significant difference between the group of ATC experts and students. In fact, the *ThetaF/AlphaP Index* of the ATC students was significantly ($p = 0.005$) higher than the ThetaF/AlphaP Index of the experts

In order to test the discriminability between the different ATM conditions, the $W_{EEG}$ index has been evaluated for each difficulty level (Easy, Medium, and Hard) and group (ATC Students and ATC Experts). Figures 7.3, 7.4 report the $W_{EEG}$ of the ATC Experts and ATC Students, respectively, calculated along the ATM scenario. In particular, the solid blue lines represent the $W_{EEG}$ index for each run (i.e., $\sim$ 3 min average), and the colored rectangles represent the $W_{EEG}$ index averaged over each difficulty level (Easy, Medium and Hard, respectively in green, yellow, and red). The figures show that for each group, the $W_{EEG}$ index was able to track the profile of the ATCO's mental workload experienced along the ATM scenario.

*Discussion and Conclusions*

The task performance analysis related to the execution of the ATM scenario provided important indications: significant difference was found between the two groups (Fig. 7.1), as the experts needed shorter time to react and provide information to the pilots than the ATC Students [$F(1, 31) = 4.7621$; $p = 0.03679$]. In terms of experienced mental workload, the results (Fig. 7.2) showed a significant difference between the ATC Experts and ATC Students [$F(1, 35) = 8.8145$;

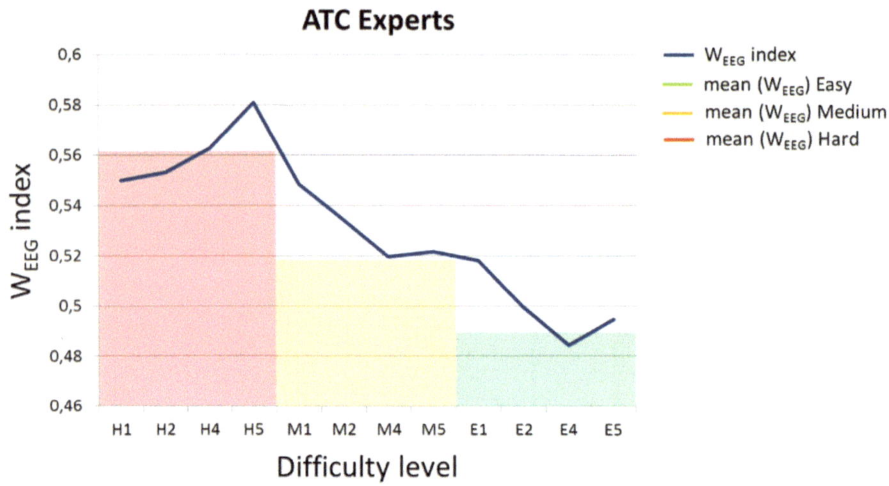


**Fig. 7.3** Representation of the $W_{EEG}$ index across the different difficulty condition (E1 ÷ E5—*green bar*, M1 ÷ M5—*yellow bar*, and H1 ÷ H5—*red bar*) and its mean value for the ATC experts

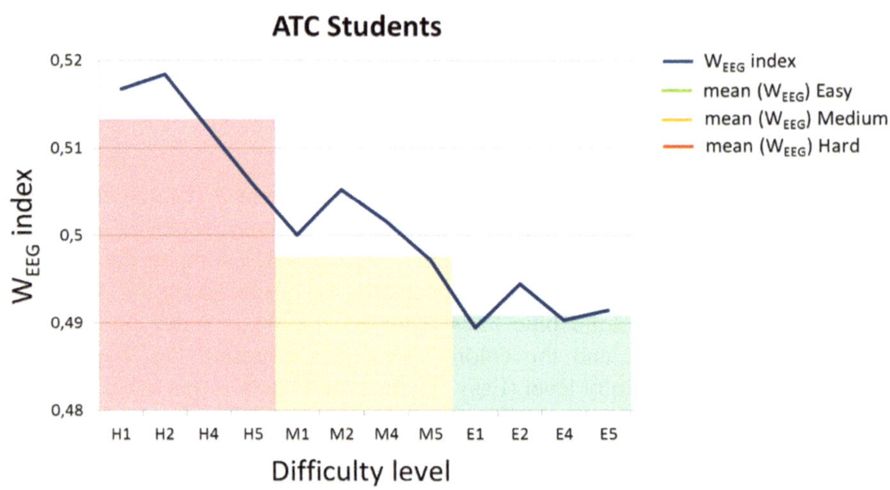


**Fig. 7.4** Representation of the $W_{EEG}$ index across the different difficulty condition (E1 ÷ E5—*green bar*, M1 ÷ M5—*yellow bar*, and H1 ÷ H5—*red bar*) and its mean value for the ATC students

$p = 0.005$]. Regarding the machine learning analysis, the asSWLDA was able to track the mental workload of the ATCOs along the execution of the ATM scenario (Figs. 7.3, 7.4), as well as to significantly differentiate the three difficulty levels proposed in the ATM scenario.

It should be underlined that, in a realistic environment, different sources of artifacts might affect the recorded neurophysiological signals, and this could produce wrong estimations and misclassifications of the physiological workload measures. ATCOs, during their operational activity, normally communicate verbally and perform several movements. Notwithstanding, the "speaking" and the other artifacts were not an issue, because the analyses have been performed (1) by filtering the frequency of the neurophysiological signals, (2) by considering only the EEG channels strictly correlated to the examined cognitive phenomenon, and (3) by using a methodology (the machine learning approach) that is able to reject automatically those features not important for—and related to—the observed cognitive phenomenon (i.e., workload).

### 7.1.2 Mental Workload and CRM Evaluation in Commercial Fligths

We simultaneously collected behavioral, psychological, peripheral, and neurophysiological data from pilot crews—*Captain* (CPT) and *First Officer* (FO)—during simulated flights at the Training Center of the Alitalia airline. The workload and cooperation conditions were modulated by inducing unexpected events during the flight. In particular, we simulated a specific electrical failure of the captain's instrumentation, in order to modify the *Pilot Flying* (PF) and *Pilot Not Flying* (PNF) roles of the crew within the flight. Such a role-exchanging contributed to create a nonstandard situation which required a higher degree of cooperation between the two pilots in order to complete the mission safely. The aim of the study was to provide objective information regarding the degree of CRM and mental workload experienced by the crews within the different flight phases by means of the pilots' neurophysiological signals.

*Participants*

Twelve commercial pilots (age: $42.8 \pm 7.1$ years old) have been recruited from the *Alitalia* airline. In particular, the six CPTs had a flight hours' total mean of $11000 \pm 2500$, while the six FOs had a flight hours' total mean of $5900 \pm 650$ (mean $\pm$ std). The flight mission was planned to include usual and unusual situations which the pilots were trained to deal with. The examined flight phases were associated with particular levels and kinds of interaction between the two pilots, that is *Takeoff*, *Cruise*, and *Landing phase*.

*Results*

The results of the ANOVA performed on the *Eye Blink Rate* (EBR) for the CPTs and for the FOs did not return any significant effects of the factors FLIGHT-PHASE and ROLE. However, a t-test between the EBR during each flight phase and the taxi returned significant differences for takeoff and landing. CPTs kept the EBR almost constant across the considered flight phases, while the CPTs showed a significant reduction of the EBR during the takeoff and landing phase. In fact, during such flight phases, the FOs had to check all the information coming from the cockpit instruments, especially during the landing. Due to the electrical failure, only the FO side was functioning, thus they had to pay lot of attention to the aircraft parameters provided by the instruments and, as consequence, the FOs showed a lower EBR than the CPTs. The *Heart Rate Index* (linear combination of the HR and the arterial pressure–HRindex) during different phases of the simulated flight revealed a significant variation of the HR according to the *within* factor FLIGHT-PHASE (F = 8.77, p = 0.00184) and to the interaction factor FLIGHT-PHASE x ROLE (F = 4.17, p = 0.031). As for the FOs, the HRindex was significantly higher during the landing with respect to the other phases. Moreover, the HRindex of the CPT and of the FO significantly differed only in the landing phase.

*Spectral Analysis*

Spectral activations at scalp level were evaluated by the *Power Spectrum Density* (PSD) of the EEG signals in the theta and alpha bands. The group's statistical spectral maps were obtained by representing on a 2D scalp model the significant t-values achieved for each of the electrodes by the statistical group analysis. T-values under the significance threshold ($p < 0.05$) were mapped in gray. The results of the statistical spectral mapping performed separately for the two groups (CPT and FO) are reported in Fig. 7.5. As for the CPT group, the only significant differences with respect to the taxi phase were found for frontal areas in the theta

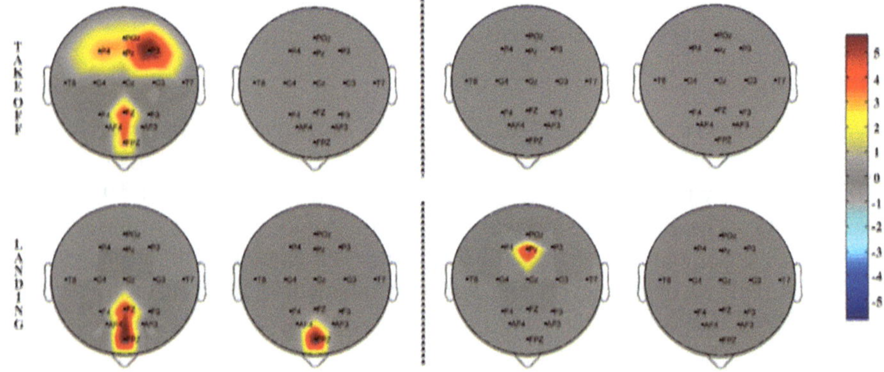


**Fig. 7.5** Group statistical spectral scalp maps computed, separately for Captains and First Officers in the theta and alpha EEG bands. Rendering of the scalp model is viewed from the above (*nose pointing* downward). The color of each pixel codes for the t-value achieved, for the related electrode, by comparing the spectral activity elicited during each flight phase with respect to the taxi phase (color scale on the *right*). All of the t-values below the statistical threshold are reported in *gray*

band during the landing phase. The FO group showed instead a significant spectral activation of the frontal and parietal areas during the takeoff phase in the theta band. During the landing phase, FOs showed frontal brain activations in the theta band (similar to what was found in the CPT group) and parietal brain activations in the alpha band.

*Crew Resource Management (CMR)*

The results of the inter-brain connectivity analysis were reported in Fig. 7.6 for a representative flight crew. All of the couples except one showed a denser pattern of interconnections linking the two brains' activities during the two cooperative flight phases, with respect to the cruise phase (no cooperative phase). Such results were confirmed in both the theta and alpha EEG bands. To statistically confirm such density modulation according to the degree of cooperation between the two pilots, we defined an index (Interconnections Density) quantifying the number of links exchanged between the two crew members. Such index was then subjected to ANOVA to evaluate its ability in discriminating the different levels of cooperation during the flight. The results of the ANOVA showed a significant effect of the main FLIGHT-PHASE *within* factor in both the theta ($F = 4.57$, $p = 0.039$) and alpha ($F = 12.186$, $p = 0.00208$) bands. Takeoff and landing were characterized by a higher number of interconnections with respect to the cruise phase, with a prevalence of connections in the landing.

*Discussion and Conclusions*

The analysis of the psychological data showed that the perceived mental workload were higher for the FOs than for the CPTs (Fig. 7.9). In fact, the higher workload perceived by the FOs might be associated with the execution of unusual actions during the landing, while the higher frustration felt by the CPTs might be put in relation with the impossibility to use their instrumentation after the electrical failure. The failure effect on the perceived workload was also confirmed by the analyses of the autonomic parameters. Moreover, a higher involvement of attentive processes

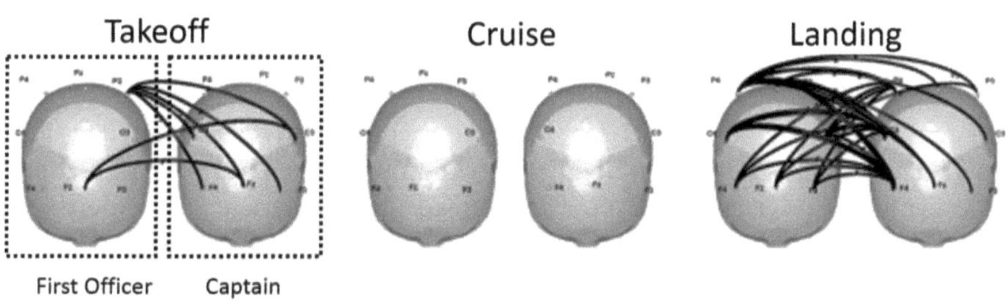


**Fig. 7.6.** Inter-brain connectivity patterns. Statistically significant connectivity patterns elicited during the takeoff, cruise, and landing phases on a representative flight crew (First Officer—FO on the *left*, and Captain—CP on the *right*). The scalp is seen from the above with the nose pointing toward the bottom of the page. Statistical connections between the pilots' brains activities are represented by *black arrows*

(reduction of EBR) and a stronger emotive engagement (increase of HRindex) for the FOs with respect to CPTs during the landing confirmed what was found in the analysis of the psychological data (Borghini et al. 2016b). Furthermore, connectivity patterns extracted for each crew showed a high number of significant information flows linking the two pilots during the most cooperative phases (takeoff and landing) of the simulated flight. Such interconnections broke down to nil during the cruise, in which the two pilots acted independently (Fig. 7.6).

The methodologies and technologies presented in this work have been extensively described and published in (Toppi et al. 2016). They were demonstrated to be employable in assessing pilot's mental workload demand and the degree of CRM states within real working environments (e.g., commercial flights). In fact, considering synchronized EEG signals of two operators as a unique dataset for inter-brain connectivity estimations, and adopting advanced techniques for the estimation of connections between brains (hyperconnectivity) will allow us to increase the knowledge about cooperation mechanisms and objectively evaluate the degree of CRM within the considered couple or team.

### *7.1.3 Military Pilots Comparison from a Cognitive Perspective*

The aim of this study was to compare two groups of military pilots, experts and novices, dealing with a simulated flight by evaluating their experienced mental workload. In this regard, the pilots were asked to perform the same mission while their neurophysiological signals (EEG, ECG, EOG) were recorded. Neurophysiological signals gathered along the flight simulations were then analyzed, and different neurometrics were computed for monitoring and getting information about the cognitive and physiological state of the pilots throughout the different flight conditions. It is well known how pilots—when facing the same flight mission—may exhibit different profiles of their mental engagement (Parasuraman and Jiang 2012): being able to quantify these cognitive differences could provide objective information for better tailoring training and expertise assessment.

*Experimental Participants*

Thirteen military pilots have been selected according to their rank and flight experience, and then have been divided in two groups (mean age $\pm$ std): 6 experts (34 $\pm$ 3) and 7 novices (26 $\pm$ 1). The average flight experience of experts consisted in more than 1450 flight hours, of which 467 on the MB339, while the novices' flight experience average was of 157 h, of which 36 on the MB339.

*Preliminary Results*

The analysis has been conducted for each pilot by calculating the *Mental Workload Index* as ratio between the frontal theta and parietal alpha PSDs within each flight phase, and then the comparison between the two groups was performed by averaged

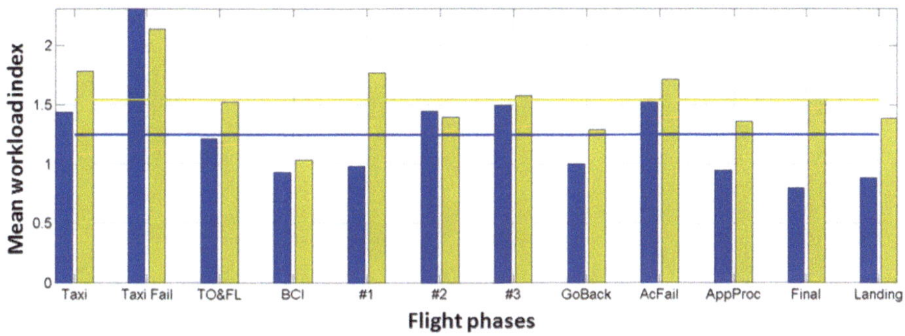


**Fig. 7.7** Trends of the experts' workload index (in *blue color*) and of the novices' workload index (in *ochre color*). The horizontal lines are the averaged index values and it is easy to note how the novices' index is greater than the novices' index

the values of such parameters. In Fig. 7.7 are shown the trends of the experts' workload index (in blue color) and of the novices' workload index (in ochre color). The horizontal lines are the mean values: it is easy to note how the novices' index was higher than the experts' one, demonstrating how novices experienced higher workload level when dealing with the same flight scenario.

*Comparison of the Cortical Maps*

Cortical maps of the differences between the experts' PSD and the novices' PSD were calculated, and only significant differences ($p < 0.05$) have been reported. In particular, brain areas showing no statistical differences were colored in gray. On the contrary, positive differences (positive t-values) were colored in red, and negative differences (negative t-values) in blue. Figure 7.8 shows two examples of cortical maps: on the top there is PSDs difference in the *theta* band during the recovery flight phase from *Nose Down* upset. The blue colors show that novices' theta PSD synchronization was higher than the experts' ones, in particular over the frontal brain areas. On the bottom of the image, it is reported the PSD difference in the *alpha* band during the resolution of the *Ac essential bus* failure. The red color shows that the desynchronization of the novices' alpha rhythm was higher than in the experts during the same task, especially over the posterior areas of the brain.

*Subjective Workload Perception*

The mean NASA-TLX and difficulty scores for the experts were 38.06 and 1.84, respectively, while, those of the novices were 55.2 and 2.34, respectively. Correlating these results with the previous analyses, it was evident how the novices sustained cognitive and physical workload levels higher than those of the experts. These workload differences were also confirmed by a subjective analysis of the external instructors, who followed the pilot's behaviors during the entire flight simulation from the remote control station. In particular, it has been reported that during the recovery from unusual flight upsets, 5 out 7 novices had applied the wrong procedures, and 6 out of 7 of them—after the resolution of the second

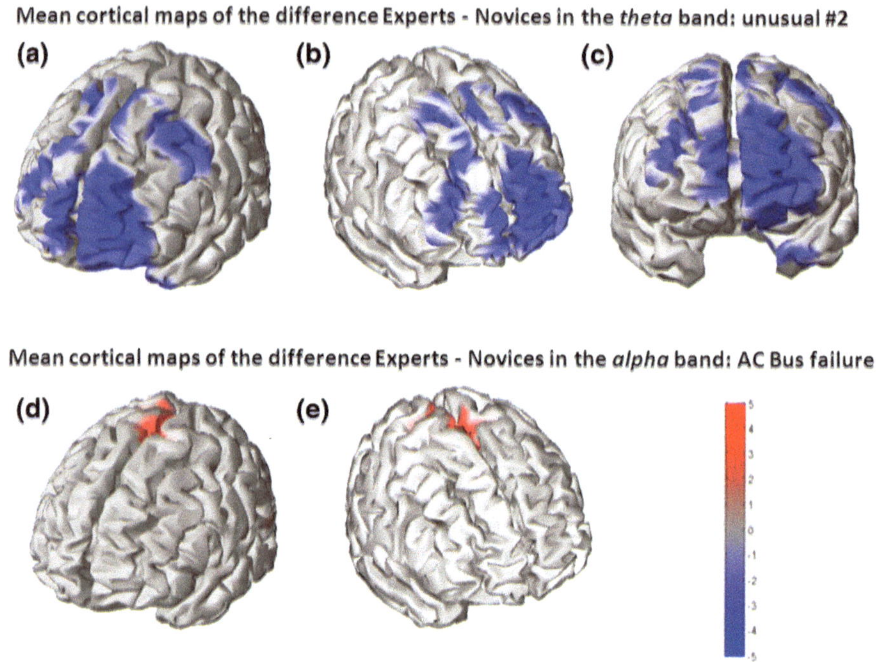


**Fig. 7.8** Cortical maps of the differences between the experts' PSD and the novices' PSD, showing only the differences statistically significance ($p < 0.05$); the gray brain areas are not significance. On the top there are different views of the cortical map of the PSD difference in the *theta* band during the flight phase of "*Nose Down*" upset recovery. **a** left lateral view, **b** right lateral view and **c** frontal view. On the bottom there are the cortical maps relative to the PSD difference in the *alpha* band during the resolution of the *Ac essential bus* failure; **d** left lateral view and **e** right lateral view

failure—have shown clear signs of stress and overload: they started breathing more frequently and deeply, sweating and asking vectoring support for returning to the base. In addition, one novice pilot kept silent and froze for almost 5 min heading toward the wrong direction, until the examiner intervened and headed the pilot back to the base and during the landing phase.

*Discussion and Conclusions*

The results showed that along the flight simulation the experts felt more relaxed than the novices and experienced lower mental workload demands.

Furthermore, the brain imaging and autonomic parameters showed how high demanding flight conditions (e.g., *Ac essential bus* failure resolution, *Nose Down* recovery) resulted more difficult for the novices than the experts, as the brain activations of the formers were higher than the latters. In fact, such conditions needed precise procedures, and the experts were probably more prone to deal with them, along with all the main actions to actuate. For the novices, instead, the situations were probably new, more complex, and demanding.

Finally, the study demonstrated that neurophysiological measures could be used in realistic environments to objectively assess the level of expertise and experienced workload level during the execution of operative tasks. Detailed descriptions and more evidences are reported in Borghini (2012b).

### 7.1.4 Mental Workload Evaluation in Real Flights

*Participants at Alenia Aermacchi*

Flights with the M311 and M346 aircrafts have been conducted by the *Flight Test Crew* of Alenia Aermacchi. The approach and the technical instrumentation have been specifically selected in terms of electrical compatibility, portability, and signal reliability. Figure 7.9 reports the setup phases of the experiments, where the pilots wore the EEG cap with the sensors under the helmet (1), the EEG system was

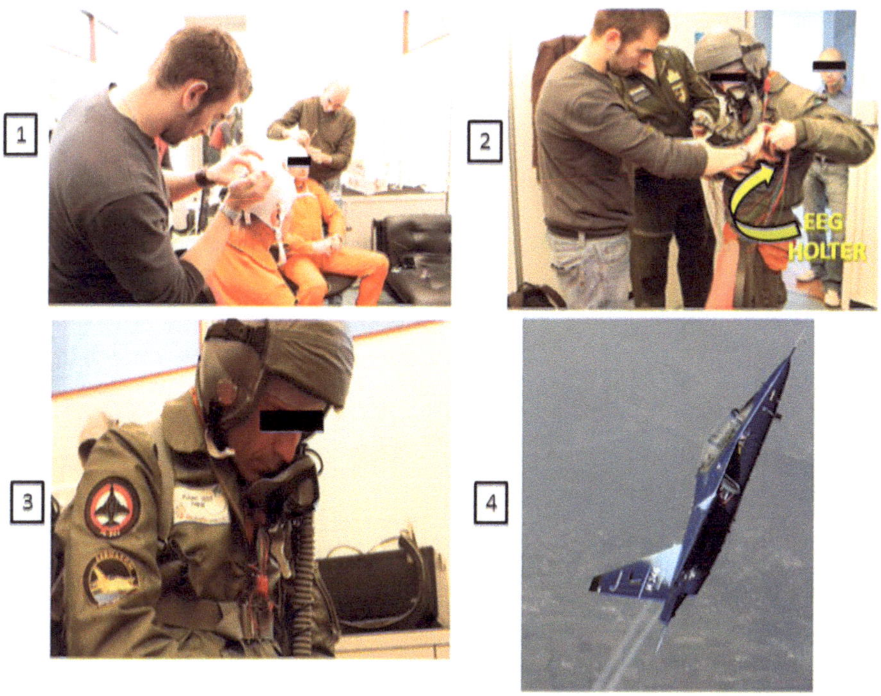


**Fig. 7.9** Experimental sessions at Alenia Aermacchi. *1* Pilots wore the EEG cap under the helmet; *2* EEG system was placed inside the flight uniform; *3* when finished, pilots went into the airplane and getting ready for taking off; *4* disorientating manoeuvres were done before the unusual attitudes

placed inside the flight uniform (2) and, when finished, the pilots went into the airplanes and got ready for taking off (3), (4) disorientating manoeuvres were done before the unusual attitudes.

The flights were presented to the pilots as a training for unusual attitudes recovery. In particular, while the test pilot kept his eyes closed, the second pilot took the plane to unusual aptitudes after few disorienting manoeuvres, and once reached the specific configuration, he gave the test pilot back the aircraft commands.

*Preliminary Results*

The preliminary results are shown in Fig. 7.10. On the left side of the figure, the training phases have been circled, and it is possible to note that, since the beginning (phase 7) the workload index (black solid line) increased within the training phases. In particular, during the *combat engagement* (phase 10) the workload index was higher than during the *standby* and *climbing* (phase 9), and after the *attack* (phase 11) it abruptly decreased. Similar discussion could be done for the unusual recoveries: the difference between the first attitude "*Nose Up*" (phase 16) and the second attitude "*Nose Down*" (phase 17) was confirmed by both the trends of the workload index and subjective pilot's rank of difficulties (red solid line).

Trends of the *heart rate* (blue solid line) and *eye blinks rate* (cyan solid line) clearly showed that along the training phases decreased, as described in paragraph 6.6 it is well known how low number of blinks means that the participant is paying more attention to what it is going on.

The heart rate (blue solid line), instead, varied during the mission and reached high values only for few conditions, *enemy interception* (phase 6), *engagement* (phase 10), *freestyle looping* (phase 13), and in the final approach for the *touch and*

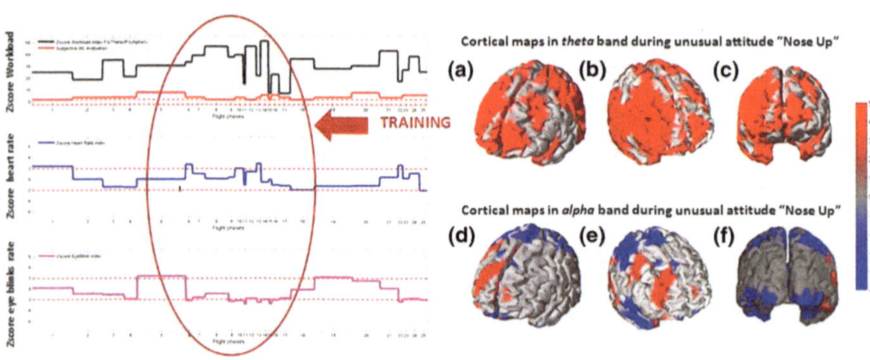


**Fig. 7.10** Indexes' trends and cortical maps referred to the experimental flight. On the left, the *black solid line* represents the mental workload, in *red* the subjective rank of difficulty compiled at the end of the experiment, in *blue* the heart rate, and in *cyan* the eye blinks rate along the different flight phases. On the right, cortical map's views in which the activated areas during the unusual "Nose Up" recovery are shown in *red colors* when the brain activity increased or in *blue* when the brain activity decreased

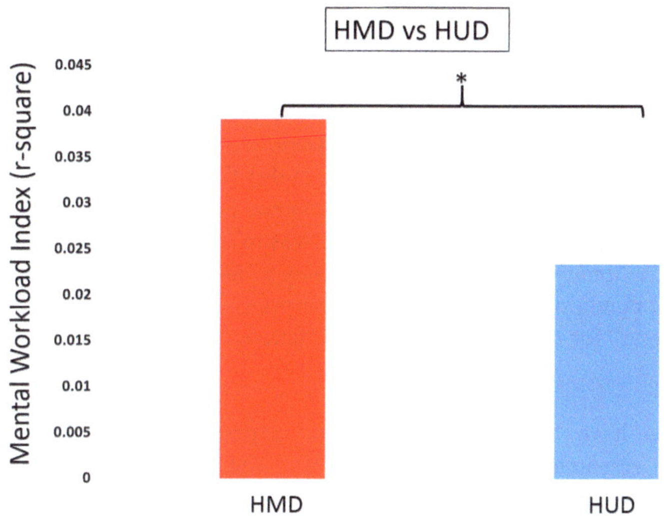


**Fig. 7.11** The HMD technology showed a higher workload demand than the HUD ($p < 10^{-5}$) over all the different visibility conditions (GVE, MVE and DVE) and technologies (FC and FC + FG). In fact, the requested MWL with the HMD was significantly higher than with the HUD

*go* (phase 22). On the right side of Fig. 7.11 there are different views of the averaged cortical maps during the recovery from the "Nose Down" recovery estimated in the *theta* band, from (a) to (c), and for *alpha* band, from (d) to (f). Red color in *theta* band denotes enhancement of brain activity over frontal areas in both brain hemispheres in comparison to rest condition, and blue color in *alpha* band indicates a brain activity decrease over posterior sites. It has to be noted that prefrontal and parietal areas are both activated in the *theta* (on the top right) and in the *alpha* band (on the bottom left).

*Discussion and Conclusions*

The results obtained in the study suggested that frontal, prefrontal, and parietal brain areas' activations were strongly correlated with the difficulty and mental demand of the flight conditions. Therefore, we can assume that neurophysiological measure could be adopted to track and evaluated the pilot's mental workload also during real flights.

Detailed descriptions and more evidences are reported in Borghini (2011).

### *7.1.5 Neuroergonomics: Avionic Technologies Testing*

The testing of avionic technologies is very important for a proper cockpit's design and for a high operational safety. Data about the pilot's mental workload might help in having a deeper comprehension of the advantages of a technology with respect to the others in terms of requested cognitive resources. Quantitative information could then support and help the comparison among the different avionic technologies in order to better identify the best choice for specific flight conditions. Based on these assumptions, a proper *Mental Workload Index* (MWL), estimated by the pilot's EEG signal, has been defined as the ratio between the frontal theta and parietal alpha brain activities (Holm 2010; Holm et al. 2009) for testing different avionic technologies during simulated flights with professional helicopter pilots in terms of required mental workload. The aim of this study was to find out which combination of the proposed avionic technology could be suggested and selected in specific operational flight conditions, in order to keep the pilot's mental workload low and within a "safe zone" for their optimal activity (Hancock et al. 1995).

*Participants*

Helicopter pilots have been recruited for the experiments performed at the AgustaWestland research Lab in Yeovil (UK). The experimental procedures involving human subjects described in this paper have been approved by the *Institutional Review Board* (IRB). All the professional pilots involved in the study (42 ± 4 years old) took part to flight simulations where different technologies have been tested and compared by estimating their mental workload from the EEG collected during the flight activities.

*Flight Scenario*

The simulator presented an AW189 helicopter's cockpit providing fully representative cockpit ergonomics: it included fully functioning side arm controllers with trim functions. The flight model provided a realistically closed control loop experience, including a representative autopilot mode behavior. The *Full Conformal* (FC) symbology consisted in the representation of important information elements that are crucial for the pilots to fly the rotorcraft like the flight path vector or the expected runway borders. The *Flight Guidance* (FG) consisted in elements that aided the pilot in maintaining the adequate trajectory. The FG has been provided as "rings in the sky" that the pilot should virtually enter during the drive. The *Synthetic Vision System* (SVS) consisted in a color-rendered terrain with grid, together with the obstacle representation using standard graphical octagons, which corresponded to the real size of the obstacles. The aim of the SVS was to increase the perception of the pilot about the surrounding environment based on a set of data stored in embedded databases. First, it has been investigated the use of *Head-Mounted Display* (HMD) and the *Head-Up Display* (HUD) when only FC or FC + FG were provided on the displays under three different visual conditions: *Good Visual Environment* (GVE), *Marginal Visual Environment* (MVE), and *Degraded Visual Environment* (DVE). Furthermore, the second phase aimed to evaluate the relative benefits of providing information about the terrain and obstacles by the SVS on the *Head-Down-Display* (HDD). The SVS has been integrated with FC (FC/SVS NO Ob) or with FC + FG and *Obstacle Radar* (RO) warning (FC + FG/SVS + RO) during flight simulations in MVE condition. Twenty-seven flight scenarios have been defined as combination of visibility, display, and symbology and they have been proposed randomly to the pilots in order to avoid habituation and expectation effects. The hypothesis was that the integration of FG or RO could reduce significantly the mental workload in different operational and visibility conditions when compared to the use of only the FC.

*Results*

Figure 7.11 reports the results of the t-test related to the MWL considering all the different visual conditions (GVE, MVE, and DVE) when the HMD (red bar) or the HUD (blue bar) was used. The results showed that the use of the HUD significantly decreased ($p < 10^{-5}$) the workload level for the accomplishment of the considered scenarios with respect to the HMD.

T-tests' analyses was also performed in order to check if the FG could reduce the MWL respect with the only FC. The results (Fig. 7.12) showed that when FC + FG information was provided (yellow bar), the experienced workload demand was significantly lower than with the only FC (green bar), both with the HMD ($p < 10^{-5}$) and the HUD ($p < 0.05$). In fact, in the scenario with only FC, the MWL was statistically higher ($p < 0.05$) than with the FC + FG.

We also performed t-tests on the NASA-TLX scores achieved during the different visibility conditions by using the two technologies (HMD and HUD). The results did not indicate significant (p = 0.87) workload changes across the different visibility conditions (GVE, MVE and DVE). Neither the workload perception of the pilots differed (p = 0.76) between the HMD and the HUD technology.

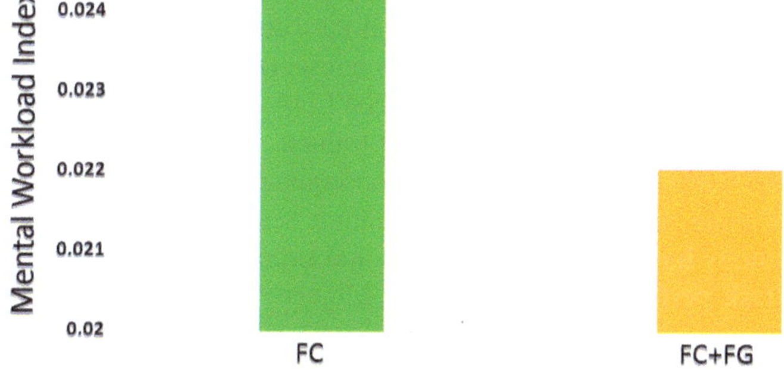


**Fig. 7.12** The integration of FG with the FC has been tested to check if the same flight conditions could require less MWL than with only FC. The t-tests showed that the FC + FG reduced significantly the MWL of the same flight scenarios both with the HMD ($p < 10^{-5}$) and the HUD ($p < 0.047$)

*Discussion and Conclusions*

The present study suggested that cognitive neurometrics are able to provide more objective and quantitative information than the subjective methodologies; detailed descriptions and more evidences reported in (Borghini et al. 2015). In fact, the results provided evidences of the benefits of HUD and HMD in reducing the workload demand (Fig. 7.15) during all flight visibility conditions ($p < 10^{-5}$). Additional flight guidance (FC + FG) statistically reduced pilots' mental workload both with the HMD ($p < 10^{-5}$) and with the HUD ($p < 0.05$), as made the same experimental conditions easier than those with only the FC (Fig. 7.12). Using the average values of the NASA-TLX scores, there have been no statistical differences ($p > 0.05$) in the subjective workload ratings both among the flight scenarios and technologies (HMD and HUD).

Considerably more physiological and subjective data would need to be recorded both under simulated and real flight conditions before standard cognitive metrics could be used as a means of providing validation in the certification process of avionic technologies. This opens up all sorts of opportunities to consider neurophysiological measures as metrics for future studies and technologies comparison.

## 7.2 Training Assessment

The aim of the study was to derive robust neurophysiological signature related to the learning progress along a training period by using biosignals such as EEG, EOG, and ECG. In order to deal with these phenomena across a training period of 3 weeks, the asSWLDA (see paragraph 6.7.1) has been used to select the most important brain features able to provide the proper resolution to track and quantify the evolution of the brain processes throughout the different sessions. The hypothesis was that, even if the EEG is not a stationary process, there could be some brain features closely related to the observed physiological phenomenon (i.e., learning process) stable enough over the time that can be used for a reliable classification of such phenomenon. In other words, if the subject is not trained, his/her brain features and activations will change across consecutive training sessions, and this will be reflected on a decreasing of the asSWLDA performance between the sessions. On the contrary, if the subject is well trained, his/her brain features and activations will be stable over time, and, as consequence, the classifier's performance will not change among consecutive training sessions. The results of the proposed methodology have also been validated and supported by the results of the behavioral data (task performance) and by the subjective data (workload perception) analyses.

*Participants*

Ten healthy volunteers (students of the *National University of Singapore*—NUS) have given their informed consent for taking part at the experiment. The local

Ethics Committee has approved the study protocol. The selection of the participants has been done accurately in order to ensure the same mental and physical state (homogeneity of the experimental sample). In addition, in order to have low sources of variances, the experimental group has been composed only by males (25 $\pm$ 3 years old).

*Experimental Protocol*

The participants have been asked to practice and to learn to execute correctly the MATB (see paragraph 3.1) for three consecutive weeks (WEEK_1, WEEK_2, and WEEK_3). In total, they have taken part in 12 training sessions (T1 $\div$ T12), with a duration of 30 min per each (Fig. 7.13). Each training session consisted in the execution of the MATB under different difficulty conditions (e.g., Easy, Medium, and Hard) proposed twice in a random sequence in order to avoid any expectation and habituation effects. To investigate the trends and changes of the EEG parameters throughout the training period (3 weeks), as signs of learning progress, the behavioral, subjective and physiological data have been analyzed in specific training sessions (Kelly and Garavan 2005), that is, the first day of WEEK_1 (T1), the last day of WEEK_1 (T5), the first repetition of the last day of WEEK_2 (T8), the first repetition of the WEEK_3 (T11) and the last repetition of the WEEK_3 (T12).

Since the participants had never done the MATB, instructions about the execution of the task have been provided on the first day of training (T1) before starting with the experiments. To be efficient, the instructional design had to be tailored to the level of the participant, in order to avoid that the effectiveness of the training was likely random (Kalyuga et al. 2003). At the end of each experimental condition of every training session, the participants have filled in the NASA-TLX questionnaire (Hart and Staveland 1988) to gather the subjective workload perception of the proposed task. In order to select the most significant brain features that are able to track and follow the learning progress of the subjects throughout the training sessions, a two-class asSWLDA regression has been used. For each participant, 10-fold-cross-validations, five considered training sessions (T1, T5, T8, T11, T12)

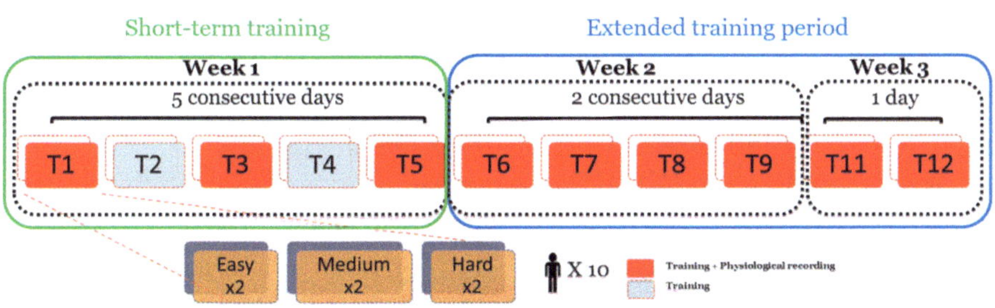


**Fig. 7.13** Extended training protocol. The participants were asked to take part in a training period of 12 sessions (T1 $\div$ T12) in which the different task conditions have been presented twice in random sequence, in order to avoid habituation and expectation effects. In the *red sessions*, the EEG, ECG and EOG signals have been recorded, while in the *gray* ones, only the behavioral and subjective data have been collected

per two task condition repetitions each session (TX-1 and TX-2, where X = 1, 5, 8, 11, 12), have been performed by calibrating the classifier with the "EASY–HARD" conditions couple of a session repetition and by testing it over the following sessions repetitions. For each couple of conditions, the *Linear Discriminant Function* has been calculated and the *Area Under Curve* (AUC) values of the *Receiver Operating Characteristic* (ROC, Bamber 1975) have been estimated. In particular, five cross-validations groups have been defined (T1, T5, T8, T11, and T12), as reported in Table 7.1: within each group, an AUC value has been calculated as average of the cross-validations considering the training session of that group (e.g., T1 for T1' group) and testing it over the consecutive ones, and vice versa. Specifically, for the first four cross-validations groups (i.e., T1, T5, T8, T11), the asSWLDA has been calibrated with the EEG dataset of each repetition of the considered session (i.e., group TX', thus TX-1 or TX-2, where X = 1, 5, 8, 11) and tested on the following sessions repetitions (i.e., TY-1 and TY-2, where Y = 5, 8, 11, 12, with Y > X). For each combination, then the roles of training and testing dataset have been inverted (e.g., considering group TX', the AUC values were computed training the classifier with TX-1/2 and testing it on TY-1/2 and vice versa). In the case of the last cross-validations group, i.e., T12', since there were not following sessions, the asSWLDA has been trained with one repetition and tested on the remaining one. In total, the number of cross—validations *(#Cross-validations)* for each one of the considered sessions has been 32 for T1, 24 for T5, 16 for T8, 8 for T11 and 2 for the last cross-validations group (T12'). The expected results have been based on the following hypotheses: (i) if the subject is well trained, the

**Table 7.1** Cross-validation scheme

| Cross-validation | Calibration dataset | Testing dataset | #Cross-validations |
|---|---|---|---|
| T1 | T1-1, T1-2 | T5-1, T5-2, T8-1, T8-2, T11-1, T11-2, T12-1, T12-2 | 32 |
| | T5-1, T5-2, T8-1, T8-2, T11-1, T11-2, T12-1, T12-2 | T1-1, T1-2 | |
| T5 | T5-1, T5-2 | T8-1, T8-2, T11-1, T11-2, T12-1, T12-2 | 24 |
| | T8-1, T8-2, T11-1, T11-2, T12-1, T12-2 | T5-1, T5-2 | |
| T8 | T8-1, T8-2 | T11-1, T11-2, T12-1, T12-2 | 16 |
| | T11-1, T11-2, T12-1, T12-2 | T8-1, T8-2 | |
| T11 | T11-1, T11-2 | T12-1, T12-2 | 8 |
| | T12-1, T12-2 | T11-1, T11-2 | |
| T12 | T12-1 | T12-2 | 2 |
| | T12-2 | T12-1 | |

AUC of such cross-validation (i.e., T1, T5, T8, T11), and the consecutive ones, will show comparable value to the *gold standard* condition (no significant difference, $p > 0.05$), that in the considered study is the last cross-validation (T12'); (ii) On the contrary, if the subject is not well trained, the AUC of the considered cross-validation will show a significant lower AUC value than the *gold standard* condition; (iii) Since the subject is trained, the brain features might remain almost stable within the same session, especially after 3 weeks of training, therefore the T12' cross-validation should show the highest AUC value.

*Preliminary Results*

The ANOVA on the task performance index highlighted significant differences [$F(4,36) = 9.92$; $p = 0.00002$] across the different training sessions (Fig. 7.14). In particular, it showed that at the end of the WEEK_1 (T5), the subjects improved significantly the level of their task performance ($p < 0.05$), and then kept it high and stable (no significant differences, $p > 0.05$ throughout the rest of the training period —that is, until the last session (T12). In fact, the Duncan's post hoc test did not find any differences, in terms of task performance, between the last day of WEEK_1 (T5) and the remaining training sessions (T8, T11 and T12).

In Fig. 7.15, the total scores of the participants' workload perceptions (NASA-TLX score) across the considered training sessions are reported. The ANOVA analysis showed a significant effect on the factor SESSION [$F(4,36) = 4.14$;

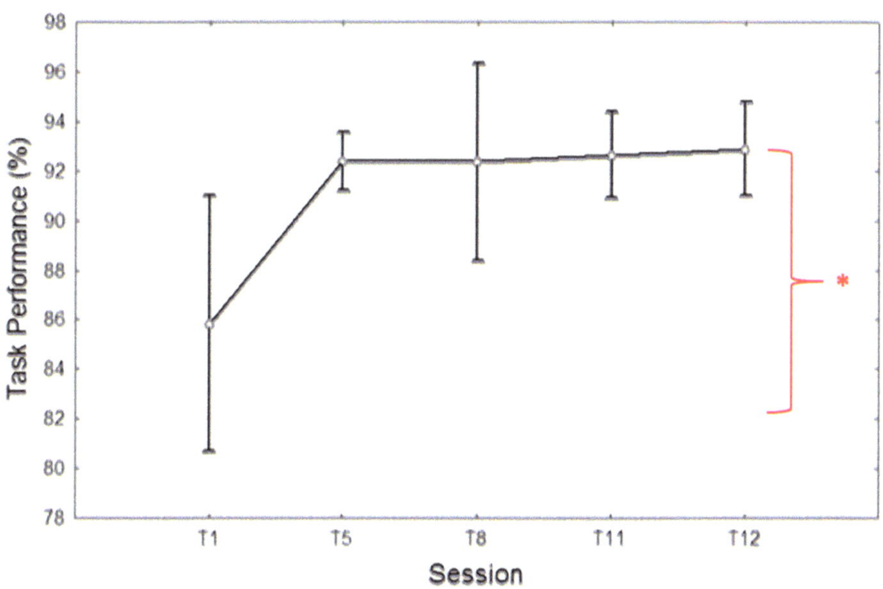


**Fig. 7.14** Task performance values over 3 weeks of training. The ANOVA showed a significant ($p < 0.05$) improvement of performance from T1 to T5 and then no differences were found between the consecutive training sessions (T5 ÷ T12). Such results indicated that since T5 the participants reached the saturation area in terms of task performance

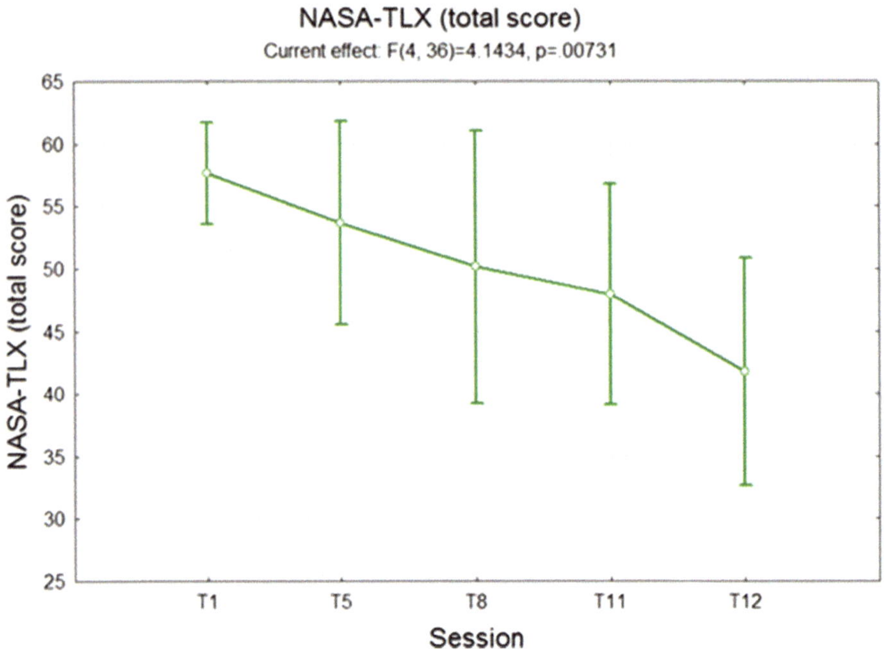


**Fig. 7.15** The figure reports the ANOVA on the NASA–TLX total scores across the considered training sessions. The results showed that the perception of the workload decreased significantly, session by session. In other words, the participants were feeling more confident and familiar with the task after each training session

p = 0.00731]. The Duncan post hoc test showed significant differences ($p < 0.03$) in the perception of the task's workload between WEEK_1 (T1 and T5) and WEEK_3 (T11 and T12). In the WEEK_2 (T8), the perception of the workload was slight lower than in the WEEK_3, but not significantly different from the last session of WEEK_1 (T5).

The ANOVA results (Fig. 7.16) highlighted significant differences [F (4,36) = 4.77; p = 0.003] between the cross-validations (T1 ÷ T12). In particular, the Duncan's post hoc test showed that the AUC value related to the first cross-validation (T1) was significantly lower (all $p < 0.02$) than those related to the others (T5, T8, T11, T12). On the contrary, from the second cross-validation (T5), no significant differences on the AUC values have been highlighted among the remained cross-validations (T8, T11, and T12).

In Fig. 7.17, the single cross-validation has been plotted with the aim of better illustrating the differences along the training sessions. In particular, the AUCs of the cross-validations by calibrating the asSWLDA on T1 (red bars), on T5 (yellow bars), on T8 (blue bars), on T11 (green bars) and on T12 (black bar) have been reported. The black dashed line indicates the trend of the averaged AUCs from T1 to the last training session (T12). As illustrated above, there was a significant

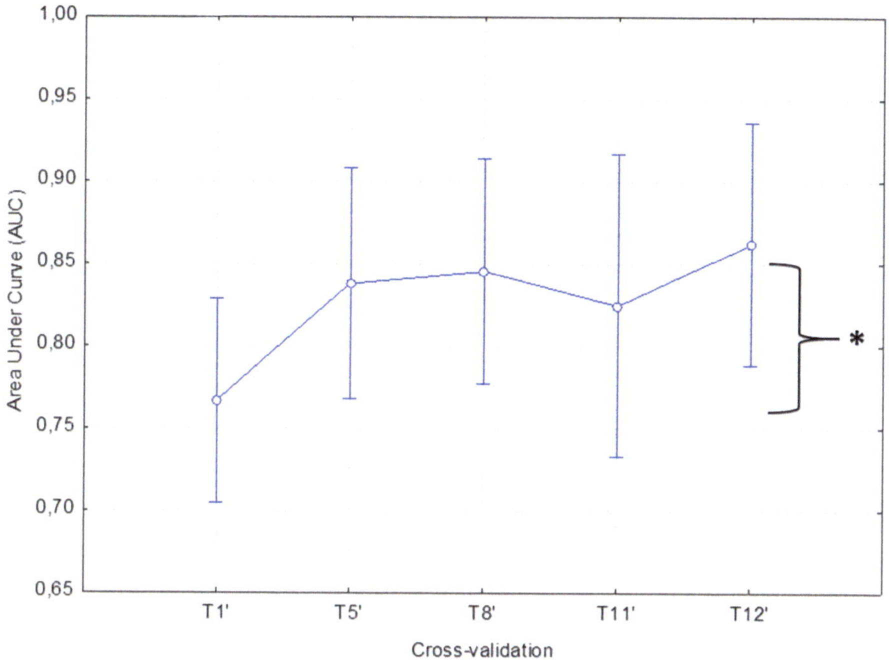


**Fig. 7.16** The ANOVA on the AUCs highlighted significant differences across the training sessions. In particular, the AUCs related to the first cross-validation (T1) was significantly lower (all $p < 0.02$) than those related to the others (T5, T8, T11, T12). On the contrary, from the second cross-validation (T5), no significant differences have been highlighted among the remained cross-validations (T8, T11, and T12)

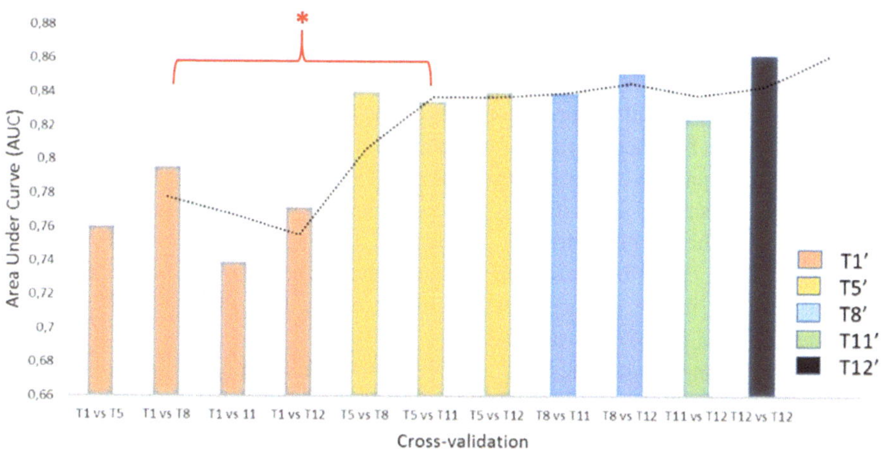


**Fig. 7.17** In the figure, the single cross-validation has been plotted with the aim of better illustrating the differences among the training sessions. In particular, the AUCs of the cross-validations by calibrating the asSWLDA on T1 (*red bars*), on T5 (*yellow bars*), on T8 (*blue bars*), on T11 (*green bars*) and on T12 (*black bar*) have been reported. The dash black line indicates the trend of the averaged AUCs and it highlights the stability (no significant differences) of the selected brain features from T5 to the last training session (T12)

increment of the AUCs from T1 to T5, and then no further differences were found among the remaining sessions.

*Discussion and Conclusions*

In this study, it was demonstrated that a machine learning approach could be used to objectively assess if and when the participants become trained, especially in terms of cognitive resources. In fact, there were no differences in calibrating the classifier with the brain features extracted from one training session rather than the others, once the participants have become trained. The analysis of the task performance (Fig. 7.14) highlighted the existence of a training effect since the last day of WEEK_1 (T5), the participants kept their performance level significantly higher ($p < 10^{-4}$) than the one that they had in the beginning of the training (T1), and no differences were found among the rest of the sessions (T5 ÷ T12). In terms of workload perception, in the last week of training (WEEK_3) the participants perceived a workload demand significantly lower ($p = 0.007$) than the one of the previous weeks (WEEK_1 and WEEK_2), but no differences have been found between WEEK_2 and WEEK_3 (Fig. 7.15). In terms of AUCs (Figs. 7.16, 7.17), the asSWLDA showed a significant increment ($p = 0.03$) from T1 to T5 and then no differences among the rest of training sessions (T5 ÷ T12). Such results suggested that from T1 to T5 the participants improved their training level with the objective to correctly accomplish the MATB and to reach high performance. As consequence, the most important brain features selected by the classifier in the T1 session could be hypothesized significantly different from those selected in T5. Successively, since in T5 no significant changes were found, in terms of classifier's performance (Figs. 7.16, 7.17), up to the end of the training period (T12), it could be reasonable to retain that the brain features remained stable over all the remaining training sessions (T5 ÷ T12). In other words, the participants became trained at the end of the first week (T5) and then no significant cognitive changes happened in the next two weeks (WEEK_2 and WEEK_3).

Two main limitations of the study have to be considered at the moment: the first is that the experimental group is just sufficient to highlight some significant statistical patterns, but needs to be increased, in order to explore if other stable neuro-electrical pattern could emerge from the analysis of the EEG signals during the training. A second limitation is the proposed task (MATB) provided by the NASA. While this is good for the analysis of the brain reaction while handling multiple tasks, it could be reasonable to retain that brain activities occurring during specific training inside an airplane cockpit, or air traffic room could be different from that elicited by MATB tasks.

Therefore, the results presented in this study have to be considered as a promising step for further studies. Similar results have been highlighted in another study, in which the actual training level of medical students while learning how to use the *DaVinci* system simulator has been evaluated by using the EEG activity of the participants (Borghini et al. 2016a).

However, the results stressed the importance of the proposed approach, because different subjects could achieve the same performance level: it would not be possible to obtain information in terms of cognitive activations (and to assert possible differences between them) just by performing the task performance analysis. In other words, by the standard analysis (task performance or self-reports), it would not be possible to find out, among the group of subjects, who are the ones requiring fewer cognitive resources—hence those showing higher cognitive spare capacity to handle with unexpected or emergency events. The results highlighted the wide applicability and general validity of our method, proposed for the training assessment in operational environments. In fact, the proposed approach could be used in many and varied contexts, like flight schools, with the aim of selecting the most skilled pilots for a certain kind of activity, or also in hospitals, for the training assessment of operators during (simulated) surgery interventions.

A more detailed description of the method has been reported in Borghini et al. (2017b), in which an improved methodology and a neurometric for the training assessment at a single-subject level have been investigated as well.

## 7.3 Expertise Estimation of Professional ATCOs

The SRK framework is still lacking of a quantitative, reliable, and validated way to be measured. Assessing the level of cognitive control of real professionals working in their own environments would enable a better comprehension of the cognitive process actually activated in the real work activities. The objective of this study was (i) to assess if it was possible to differentiate the three degrees of cognitive controls proposed by the SRK model by the analysis of the Air Traffic Controllers' brain activity, and (ii) to define an objective method to overcome the actual limitations in measuring and quantifying the above-mentioned SRK cognitive behaviors. In particular, the goals are: (i) to find out if the brain features are able to define a metric meant to discriminate the SRK level, independently from those used for the mental workload evaluation; (ii) to compare ATC Experts and ATC Students, in terms of SRK behaviors, from a neurophysiological point of view. First, a literature review has been done specifically to find out those brain features mainly correlated to the cognitive processes involved in the different SRK cognitive behaviors. Among all the possible brain features, four kinds of features have been initially selected and tested: two of them are the frontal theta and the parietal alpha EEG rhythms (see Chaps. 3, 4), because of their correlation with working memory, information processing, and decision-making. The other two selected brain features are the parietal theta and frontal alpha EEG rhythms. The first is correlated to the procedural memory and the latter is correlated to the attention level (see following sections). Then, the asSWLDA (see paragraph 6.7.1) has been calibrated with the selected brain features, in order to characterize and discriminate the different cognitive control levels reached by professional Controllers during the execution of realistic ATC activities.

*Participants*

Thirty-seven professional ATCOs from the *École Nationale de l'Aviation Civile* (ENAC) of Toulouse (France) have been involved in this study. They have been selected in order to have homogeneous experimental groups in terms of age and expertise. In particular, two groups of participants have been defined, a group of ATC Experts (40.41 ± 5.54) and a group of ATC Students (23 ± 1.95).

*ATM Scenario*

The ATCOs have been asked to perform an ATM simulation using a research simulator hosted at ENAC (see paragraph 3.2). Moreover, two pseudo-pilots have attended the experiment: the latter have interacted with the ATCOs with the aim of simulating and reproducing real communications, modulating the SRK events. The ATM scenario enclosed different levels of difficulty. The traffic complexity has been modulated in terms of number of aircrafts in the controlled sector, number of conflicts and geometry complexity.

A *Subject Matter Expert* (SME) from the *Ente Nazionale di Assistenza al Volo* (ENAV, Rome, Italy) has been involved in the experiment, in order to create realistic and not disruptive SRK events during the simulation (to be limited in time, thus not changing the realism of the scenario). The events designed in the air traffic sample scenario represented an attempt to induce ATCOs' behaviors, during usual air traffic conditions, associated with the SRK levels. In particular, six SRK events have been generated: two for each level of cognitive control behavior (Fig. 7.18). The events have not been inserted in the hard part of the ATM scenarios, since the introduction of external events could increase the difficulty to an unacceptable level, generating too much disruption on the controlling activity and possibly invalidating part of the recording and of the task's realism. The design of the SRK events has been the following. The *Skill* events were basically interactions with the interface. The *Rule* events are mainly control-tasks and conflicts-resolutions. Finally, the *Knowledge* events represented unfamiliar and unusual situations: this uncertainty led the Controllers to require time to analyze the situation and to find out the proper

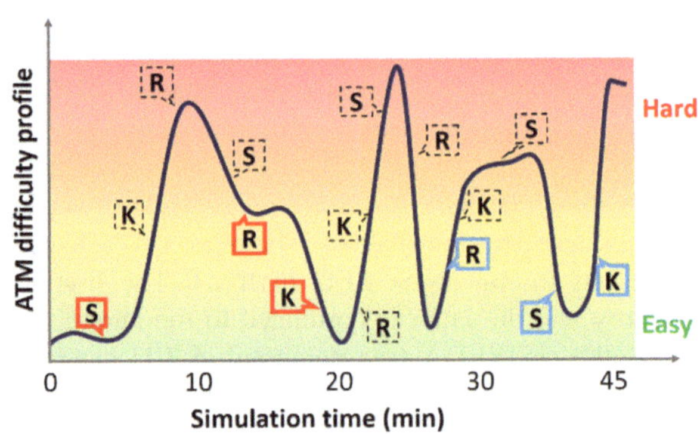


**Fig. 7.18** ATM simulation time-line as a function of traffic complexity, showing SRK events. The simulation lasted 45 min and, within it, the triplets of SRK events have been inserted within coherent difficulty conditions (*red*, *blue* and *black squares*)

**Table 7.2** Short description of the SRK events

| Name of the event | Type of event |
|---|---|
| | *Skill based* |
| Distance | Controllers are asked to measure the distance between two aircraft using the validate tool |
| Display FPs | The SME expert triggered the event giving to the controller a paper with the instruction to visualize the Flight Plan (FPL) trajectory of each aircraft assumed. Controller have to use the pie menu and select the route option for each aircraft |
| | *Rule based* |
| Conflict | Controllers have to detect and solve a conflict, using the pie menu to assign new altitudes and headings |
| | *Knowledge based* |
| Deviation | Controllers are asked to face with unexpected situation, in which an aircraft changes its route respect to the cleared FP. Controllers have to detect this manoeuvre and check the a/c flight plan on the HMI, to evaluate if it is correct or if something is going wrong |
| UFO | Controllers are asked to verify an unknown traffic detected by the Traffic Collision Avoidance System (TCAS), to avoid a mid air collision. After the avoidance manoeuvre pilots report a different FL respect to what provided by the HMI |

procedure to cope with the unexpected event. A summary of the SRK events is reported in Table 7.2.

*Preliminary Results*

As mentioned above, the asSWLDA was calibrated by using one triplet (S1, R1, K1) and the discrimination accuracy (AUC) values were calculated by testing it on the remaining triplets (S2, R2, K2) and *vice versa*. Referring on the results reported in the previous sections, the parietal theta and frontal alpha EEG rhythms were defined by the frequency domain in which the classification model had to select the most significant features. The t-tests ($\alpha = 0.05$) showed that all the measured AUC distributions were significantly higher (Fig. 7.20) with respect to the random AUC distributions (all $p < 10^{-5}$). The tests showed that, as reported for the ATC Experts, also for the ATC Students all the measured AUC distributions were significantly higher (Fig. 7.20) in comparison to the random AUC distributions (all $p < 10^{-3}$).

*Comparison between ATC Experts and Students*

In Fig. 7.21 the AUC values of the two ATCO groups are reported. The ATCO Experts (blue bar) exhibited a significant higher ($p < 0.05$) discrimination accuracy (AUC = 0.8) with respect to the ATCO Students (red bar, AUC = 0.72). The trend may indicate how the cognitive control behaviors were better discriminable for Experts ATCOs than for students.

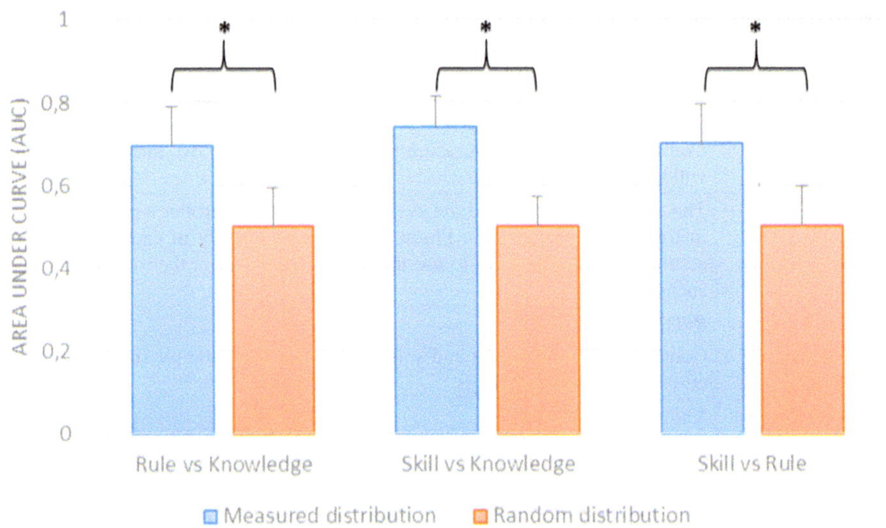


**Fig. 7.19** Error bars (CI = 0.95) related to the measured AUC distributions and the random AUC distributions, achieved by the ATC experts, referred to the discrimination accuracy between the three couples of conditions (S vs. R, S vs. K, R vs. K)

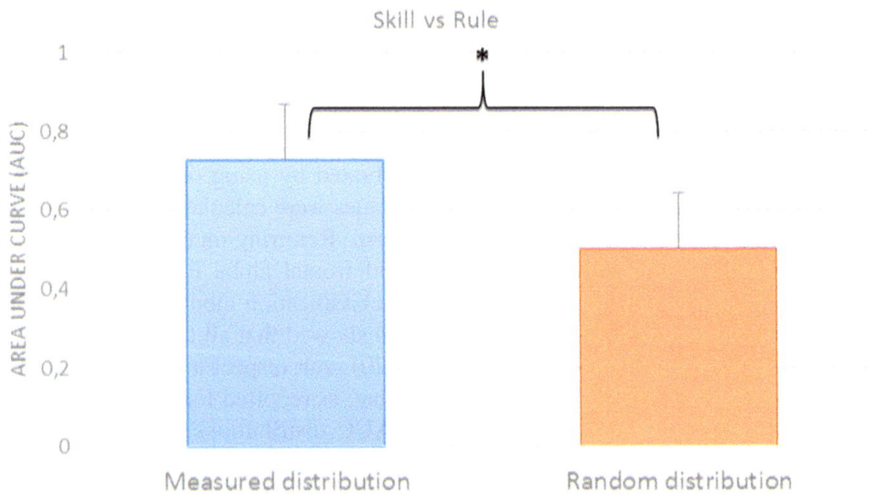


**Fig. 7.20** Error bars (CI = 0.95) related to the Measured AUC distributions and the random AUC distributions, achieved by the ATC students, referred to the discrimination accuracy between the skill and rule conditions

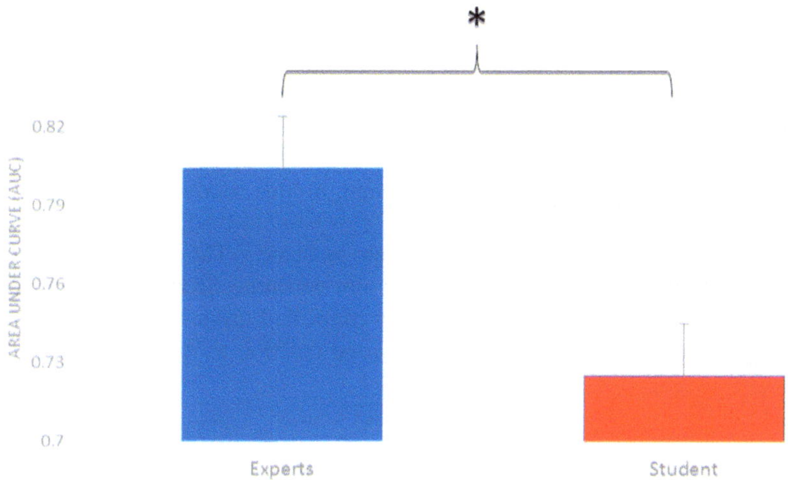


**Fig. 7.21** Error bars (CI = 0.95) related to the discrimination accuracy of the skill versus rule condition achieved by the ATC Experts (*blue bar*) and ATC students (*red bar*), referred to the discrimination accuracy between the skill and rule conditions

*Discussion and Conclusions*

Several studies tried to model the different information processing procedures based on Skill, Rule, or Knowledge control cognitive levels from a human factor point of view. Despite such studies, there are not evidences of these control cognitive levels being taken into account from a neurophysiological point of view (for instance, by considering the EEG variations between the different S-R-K levels). The selected brain features for the definition of the metric for the SRK discrimination were the parietal theta and frontal alpha EEG rhythms. The former is correlated to the procedural memory: the more is the memory consolidation, the more will be its synchronization over the parietal brain areas. The memory consolidation is correlated to the attention level, and it has been demonstrated that: the more the attention is, the higher the desynchronization of the alpha activity over the frontal brain areas is. Those brain features were used to calibrate the asSWLDA classification algorithm, and to test it about the possibility to differentiate the three cognitive control levels in real time, during the execution of a realistic ATM scenario executed by professional ATCOs. The asSWLDA was able to discriminate significantly the SRK conditions for the ATC Experts ($p < 10^{-5}$), and the S-R conditions for the ATC Students ($p < 10^{-3}$), and all the measured AUC distributions were significantly higher (Fig. 7.19 and 7.20) than the random AUC distributions (all $p < 10^{-3}$).

To sum up, in this study it has been demonstrated that it was possible to assess the ATCO's cognitive control level (SRK) with high reliability, by using a specific metric estimated from the brain activity. The proposed method has been proven to be reliable in realistic ATM settings, since the results showed that

(1) The proposed metric (parietal theta and frontal alpha) should be a valid solution to investigate and objectively analyze the different cognitive control behaviors under which the user is performing a task;
(2) The asSWLDA was able to significantly discriminate the SRK conditions for both the groups of ATCOs.

Detailed descriptions and more evidences found in this study are reported in (Borghini et al. 2017a).

Due to the need of inserting SRK events into realistic ATM tasks, it was not possible to create "pure" SRK behaviors. The aim of future SRK studies will be then to design experimental protocols by using controlled tasks, in order to better simulate and induce pure skill, rule and knowledge reaction to the participants.

## References

Bamber D (1975) The area above the ordinal dominance graph and the area below the receiver operating characteristic graph. J Math Psychol 12:387–415. doi:10.1016/0022-2496(75)90001-2

Bekhet AK, Zauszniewski JA (2012) Methodological triangulation: an approach to understanding data. Nurse Res 20:40–43. doi:10.7748/nr2012.11.20.2.40.c9442

Borghini G (2011) Brainshield: HREEG study of perceived pilot mental workload. Ital J Aerosp, Med 1

Borghini G, Isabella R, Vecchiato G, Toppi J, Astolfi L, Caltagirone C, Babiloni F (2012) A flight history from a cognitive point of view: novices versus experts. Ital J Aerosp Med 5:34–47

Borghini G, Aricò P, Di Flumeri G, Salinari S, Colosimo A, Bonelli S, Napoletano L, Ferreira A, Babiloni F (2015) Avionic technology testing by using a cognitive neurometric index: a study with professional helicopter pilots. In: IEEE annual international conference on the Engineering in Medicine and Biology Society (EMBC), 2015. pp 6182–6185. doi:10.1109/EMBC.2015.7319804

Borghini G, Aricò P, Di Flumeri G, Colosimo A, Storti SF, Menegaz G, Fiorini P, Babiloni F (2016a) Neurophysiological measures for users' training objective assessment during simulated robot-assisted laparoscopic surgery. In: IEEE annual international conference on the Engineering in Medicine and Biology Society (EMBC), 2016

Borghini G, Aricò P, Graziani I, Salinari S, Sun Y, Taya F, Bezerianos A, Thakor NV, Babiloni F (2016b) Quantitative assessment of the training improvement in a motor-cognitive task by using EEG, ECG and EOG signals. Brain Topogr 29:149–161. doi:10.1007/s10548-015-0425-7

Borghini G, Aricò P, Di Flumeri G, Cartocci G, Colosimo A, Bonelli S, Golfetti A, Imbert J-P, Granger G, Benhacene R, Pozzi S, Babiloni F (2017a) EEG-based cognitive control behaviour assessment: an eco-logical study with professional air traffic controllers. Sci Rep—Nat (in press)

Borghini G, Arico P, Di Flumeri G, Sciaraffa N, Colosimo A, Herrero Ezquerro MT, Bezerianos A, Thakor N, Babiloni F (2017b) A new perspective for the training assessment: machine-learning-based neurometric for augmented user's evaluation

Hancock PA, Williams G, Manning CM (1995) Influence of task demand characteristics on workload and performance. Int J Aviat Psychol 5:63–86. doi:10.1207/s15327108ijap0501_5

Hart SG, Staveland LE (1988) Development of NASA-TLX (Task Load Index): results of empirical and theoretical research, in: human mental workload. North-Holland, pp 139–183

Holm A (2010) Developing neurophysiological metrics for the assessment of mental workload and the functional state of the brain. Dr. Diss. Degree Dr. Philos. Aalto University School of Science and Technology, Espoo, Finland

Holm A, Lukander K, Korpela J, Sallinen M, Müller KMI (2009) Estimating brain load from the EEG. Sci World J 9:639–651. doi:10.1100/tsw.2009.83

Jick TD (1979) Mixing qualitative and quantitative methods: triangulation in action. Adm Sci Q 24:602–611. doi:10.2307/2392366

Kalyuga S, Ayres P, Chandler P, Sweller J (2003) The expertise reversal effect. Educ Psychol 38:23–31. doi:10.1207/S15326985EP3801_4

Kelly AMC, Garavan H (2005) Human functional neuroimaging of brain changes associated with practice. Cereb Cortex 1991(15):1089–1102. doi:10.1093/cercor/bhi005

Parasuraman R, Jiang Y (2012) Individual differences in cognition, affect, and performance: behavioral, neuroimaging, and molecular genetic approaches. NeuroImage 59:70–82. doi:10.1016/j.neuroimage.2011.04.040

Toppi J, Borghini G, Petti M, He EJ, Giusti VD, He B, Astolfi L, Babiloni F (2016) Investigating cooperative behavior in ecological settings: an EEG hyperscanning study. PLoS ONE 11: e0154236. doi:10.1371/journal.pone.0154236

# Chapter 8
# General Conclusions

The present book aimed to provide state-of-the-art overviews concerning the most important and studied mental states in aviation. In particular, definitions of the workload, cognitive control behavior, and training assessment have been presented.

Objective investigations, deeper comprehension, and evaluation of such cognitive phenomena can avoid and limit the occurrence of errors and improve the level of safety within operational environments.

In fact, most of the standard methodologies rely on self-reported questionnaires (operator-dependent) or task performance evaluation (no information regarding the request of cognitive resources for dealing with the considered task).

Futhermore, it has been widely demonstrated how most of the accidents in operational contexts are due to human error mainly caused by operator's mental state impairment.

In order to deal with such limitation, several experimental protocols have been designed both in controlled settings (Lab) and in realistic environments (ATC control room, flight simulators, jet-fighters) by recruiting students and professional personnel (air traffic controllers and pilots). The studies aimed to define and validate neurometrics and methodologies for objectively assess the operator's mental state during the execution of tasks.

A modified version of the SWLDA algorithm was developed and tested in such contexts: the *automatic stop StepWise Linear Discriminant Analysis*—asSWLDA (Aricò et al. 2015a). This algorithm is able to include, in the classification model, only those features strictly related to the considered cognitive phenomenon, thus mitigating both the under- and overfitting phenomenon. In addition, even without recalibrating the algorithm within a month, the asSWLDA does not suffer any significant decrease in terms of discrimination accuracy. Another important advantage, with respect to the standard SWLDA implementation, is that the as SWLDA uses a significant lower number of EEG channels (among the available ones) to select the most significant features necessary to track and quantify the considered cognitive phenomenon. Consequently, the advantages of the asSWLDA



G. Borghini et al., *Industrial Neuroscience in Aviation*,
Biosystems & Biorobotics 18, DOI 10.1007/978-3-319-58598-7_8

are very practical since there is a significant reduction of both the number of sensors (EEG channels)—so, the invasiveness of the recording system—and the time needed to run the experiments (no calibration up to a month). In this way, the proposed method could be easily employed in working operational environments.

Furthermore, such characteristics make the proposed algorithm and related methodologies very useful to assess the learning progress of the user across a specific training period, and the cognitive control behavior under which the user is executing the considered tasks. In particular, as described in the paragraph 3.3, the main limitations in training assessments are the lack of objective information regarding the cognitive activations *within* and *between* consecutive training sessions. The results demonstrated that the proposed methodologies could overcome such limitations and provide important information to better evaluate the user's learning progress, and efficiently tailor the training programs.

Concerning the evaluation of cognitive control behaviors, no neurophysiological metrics have been proposed in the scientific literature. During the selection of personnel, it might be important to know how the users can perform the proposed task (Broach 2013). If they can achieve high performance (100%), it will not mean they can do it automatically or by involving just few brain resources. By the standard evaluation methods, it is not possible to have this kind of information. The results of this study, explained in detail in (Borghini et al. 2017a) demonstrated that the proposed neurophysiological metric can be used to discriminate the cognitive control behavior under which the user is accomplishing the task. Therefore, the users' evaluation can be done more accurately.

To sum up, the works illustrated in this book stressed the importance of combining multimodal data and signals (i.e., brain, heart, and ocular activity) in order to overcome current limitations in assessing user's performance while dealing with and accomplishing specific tasks. Specifically, the developed methodologies could supply quantitative and objective information regarding the participants' mental workload, cognitive control behavior, and training level, in order to support the instructor for a better management and tailoring of the training schedule.

These evidences represent a promising step forward in the analysis of human behaviors, and demonstrate the possibility of developing a device that is able to evaluate in real time the operators' cognitive engagement during the working activities in operational environments. This tool could be used as well to improve complex systems' designs and to enhance the capability to anticipate errors: for instance, the *Realistic Interface Design* (RID, Pawlak et al. 1996; Vicente and Rasmussen 1992) uses the SRK framework as an analytical tool for the design of interfaces in complex sociotechnical, real-time, and dynamic systems. Another possible use might be an online tool for triggering *Adaptive Automations* (Byrne and Parasuraman 1996; Parasuraman et al. 1996; Uebbing-Rumke et al. 2012), where the systems behaves depending upon the operator's current level of cognitive control.

In particular, this theory could be applied to better understand how to set AA solutions to optimize human performance. In fact, as described previously, AA-based systems could reduce the operator's engagement so much that an *underload* situation could arise and, as a consequence, a reduction of task performance will be obtained. In the same way, if the LOA is improperly set to a too low value (i.e., the system control is mainly manual), the manual task demands could increase so much to induce an *overload* condition (Parasuraman et al. 1992). From this perspective, it is crucial to have a reliable estimation of the operator's mental status along the execution of the task to preserve a proper level of cognitive engagement, maintain optimal levels of performance, and reduce the risk of errors commission. The authors of the book have already demonstrated how it was possible to reliably evaluate the ATCO's mental workload in realistic context and trigger AAs when the workload overcame a specific threshold in order to support and keep the ATCO's workload within a safe range (Aricò et al. 2016).

## References

Aricò P, Borghini G, Di Flumeri G, Babiloni F (2015) Metodo per stimare uno stato mentale, in particolare un carico di lavoro, e relativo apparato (A Method for the estimation of mental state, in particular of the mental workload and its device). P1108IT00

Aricò P, Borghini G, Di Flumeri G, Colosimo A, Bonelli S, Golfetti A, Pozzi S, Imbert J.-P, Granger G, Benhacene R, Babiloni F (2016) Adaptive Automation triggered by EEG-based mental workload index: a passive brain-computer interface application in realistic air traffic control environment. Front. Hum. Neurosci 539. doi:10.3389/fnhum.2016.00539

Borghini G, Aricò P, Di Flumeri G, Cartocci G, Colosimo A, Bonelli S, Golfetti A, Imbert J.-P, Granger G, Benhacene R, Pozzi S, Babiloni F (2017). EEG-based cognitive control behaviour assessment: an eco-logical study with professional air traffic controllers. Sci. Rep. - Nat. In press

Broach D (2013) Selection of the next generation of air traffic control specialists: aptitude requirements for the air traffic control tower cab in 2018. Aerospace Medicine Technical Reports No. DOT/FAA/AM-13/5. FAA Office of Aerospace Medicine Civil Aerospace Medical Institute

Byrne EA, Parasuraman R (1996) Psychophysiology and adaptive automation. Biol Psychol, Psychophysiology of Workload 42:249–268. doi:10.1016/0301-0511(95)05161-9

Parasuraman R, Bahri T, Deaton JE, Morrison JG, Barnes M (1992) Theory and Design of Adaptive Automation in Aviation Systems

Parasuraman R, Mouloua M, Molloy R (1996) Effects of adaptive task allocation on monitoring of automated systems. Hum Factors 38:665–679

Pawlak W, Brinton C, Crouch K, Lancaster K, (1996) A framework for the evaluation of air traffic control complexity. In: Guidance, navigation, and control conference. American Institute of Aeronautics and Astronautics

Uebbing-Rumke M, Gürlük H, Schulze-Kissing D (2012) Adaptive automation support for time-based operations in ATC. IOS Press

Vicente KJ, Rasmussen J (1992) Ecological interface design: theoretical foundations. IEEE Trans Syst Man Cybern 22:589–606. doi:10.1109/21.156574

MIX
Papier aus verantwortungsvollen Quellen
Paper from responsible sources
FSC® C105338

If you have any concerns about our products,
you can contact us on
**ProductSafety@springernature.com**

In case Publisher is established outside the EU,
the EU authorized representative is:
**Springer Nature Customer Service Center GmbH**
**Europaplatz 3, 69115 Heidelberg, Germany**

Printed by Libri Plureos GmbH
in Hamburg, Germany